MATHEMATICAL FOUNDATIONS OF

Statistical Mechanics

BY A. I. KHINCHIN

TRANSLATED FROM THE RUSSIAN BY G. GAMOW

Dover Publications, Inc.

NEW YORK

PRINTED IN THE UNITED STATES OF AMERICA
THE WILLIAM BYRD PRESS, INC., RICHMOND, VIRGINIA

LC: # 49-9707

CONTENTS

PREFACE

Statistical mechanics presents two fundamental problems for mathematics: (1) the so-called ergodic problem, that is the problem of a rigorous justification of replacement of time-averages by space (phase)-averages; (2) the problem of the creation of an analytic apparatus for the construction of asymptotic formulas. In order to become familiar with these two groups of problems, a mathematician usually has to overcome several difficulties. For understandable reasons, the books on physics do not pay much attention to the logical foundation of statistical mechanics, and a great majority of them are entirely unsatisfactory from a mathematical standpoint, not only because of a non-rigorous mathematical discussion (here a mathematician would usually be able to put things in order by himself), but mainly because of the almost complete absence of a precise formulation of the mathematical problems which occur in statistical mechanics.

In the books on physics the formulation of the fundamental notions of the theory of probability as a rule is several decades behind the present scientific level, and the analytic apparatus of the theory of probability, mainly its limit theorems, which could be used to establish rigorously the formulas of statistical mechanics without any complicated special machinery, is completely ignored.

The present book considers as its main task to make the reader familiar with the mathematical treatment of statistical mechanics on the basis of modern concepts of the theory of probability and a maximum utilization of its analytic apparatus. The book is written, above all, for the mathematician, and its purpose is to introduce him to the problems of statistical mechanics in an atmosphere of logical precision, outside of which he cannot assimilate and work, and which, unfortunately, is lacking in the existing physical expositions.

The only essentially new material in this book consists in the systematic use of limit theorems of the theory of probability for rigorous proofs of asymptotic formulas without any special analytic apparatus. The few existing expositions which intended to give a rigorous proof to these formulas, were forced to use for this purpose special, rather cumbersome, mathematical machinery. We hope, however, that our exposition of several other questions (the ergodic problem, properties of entropy, intramolecular correlation, etc.) can claim to be new to a certain extent, at least in some of its parts.

CHAPTER I

INTRODUCTION

1. A brief historical sketch. After the molecular theory of the structure of matter attained a predominant role in physics, the appearance of new statistical (or probabilistic) methods of investigation in physical theories became unavoidable. From this new point of view each portion of matter (solid, liquid, or gaseous) was considered as a collection of a large number of very small particles. Very little was known about the nature of these particles except that their number was extremely large, that in a homogeneous material these particles had the same properties, and that these particles were in a certain kind of interaction. The dimensions and structure of the particles, as well as the laws of the interaction could be determined only hypothetically.

Under such conditions the usual mathematical methods of investigation of physical theories naturally remained completely powerless. For instance, it was impossible to expect to master such problems by means of the apparatus of differential equations. Even if the structure of the particles and the laws of their interaction were known, their exceedingly large number would have presented an insurmountable obstacle to the study of their motions by such methods of differential equations as are used in mechanics. Other methods had to be introduced, for which the large number of interacting particles, instead of being an obstacle, would become a stimulus for a systematic study of physical bodies consisting of these particles. On the other hand, the new methods should be such that a lack of information concerning the nature of the particles, their structure, and the character of their interaction, would not restrict the efficiency of these methods.

All these requirements are satisfied best by the methods of the theory of probability. This science has for its main task the study of group phenomena, that is, such phenomena as occur in collections of a large number of objects of essentially the same kind. The main purpose of this investigation is the discovery of such general laws as are implied by the gross character of the phenomena and depend comparatively little on the nature of the individual objects. It is clear that the well-known trends of the theory of probability fit in the best possible way the aforementioned special demands of the molecular-physical theories. Thus, as a matter of principle, there was no doubt that statistical methods should become the most important mathematical tool in the construction of new physical theories; if there existed any disagreement at all, it concerned only the form and the domain of application of these methods.

In the first investigations (Maxwell, Boltzmann) these applications of statistical methods were not of a systematical character. Fairly vague and somewhat timid probabilistic arguments do not pretend here to be the fundamental basis, and play approximately the same role as purely mechanical considerations. Two features are characteristic of this primary period. First, far reaching hypotheses are made concerning the structure and the laws of interaction between the particles; usually the particles are represented as elastic spheres, the laws of collision of which are used in an essential way for the construction of the theory. Secondly, the notions of the theory of probability do not appear in a precise form and are not free from a certain amount of confusion which often discredits the mathematical arguments by making them either void of any content or even definitely incorrect. The limit theorems of the theory of probability do not find any application as yet. The mathematical level of all these investigations is quite low, and the most important mathematical problems which are encountered in this new domain of application do not yet appear in a precise form.[1]

It should be observed, however, that the tendency to restrict

[1]An excellent critical analysis of this first period is found in a well-known work by P. and T. Ehrenfest which appeared in vol. IV of the Encyclopaedie der Mathematischen Wissenschaften and which played a considerable role in the development of the mathematical foundations of statistical mechanics.

the role of statistical methods by introducing purely mechanical considerations, (from various hypotheses concerning the laws of interaction of particles), is not restricted to the past. This tendency is clearly present in many modern investigations. According to a historically accepted terminology, such investigations are considered to belong to the kinetic theory of matter, as distinct from the statistical mechanics which tries to reduce such hypotheses to a minimum by using statistical methods as much as possible. Each of these two tendencies has its own advantages. For instance, the kinetic theory is indispensable when dealing with problems concerning the motion of separate particles (number of collisions, problems concerning the study of systems of special kinds, mono-atomic ideal gas); the methods of the kinetic theory are also often preferable, because they give a treatment of the phenomena which is simpler mathematically and more detailed. But in questions concerning the theoretical foundation of general laws valid for a great variety of systems, the kinetic theory naturally becomes sometimes powerless and has to be replaced by a theory which makes as few special hypotheses as possible concerning the nature of the particles. In particular, it was precisely the necessity of a statistical foundation for the general laws of thermodynamics that produced trends which found their expression in the construction of statistical mechanics. To avoid making any special hypotheses about the nature of the particles it became necessary in establishing a statistical foundation to develop laws which had to be valid no matter what was the nature of these particles (within quite wide limitations).

The first systematic exposition of the foundations of statistical mechanics, with fairly far developed applications to thermodynamics and some other physical theories, was given in Gibbs' well-known book.[2] Besides the above mentioned tendency not to make any hypotheses about the nature of particles the following are characteristic of the exposition of Gibbs.

---

[2]W. Gibbs, "Elementary principles of statistical mechanics," Yale University Press, 1902.

(1) A precise introduction of the notion of probability, which is given here a purely mechanical definition, is lacking with the resulting questionable logical precision of all arguments of statistical character.
(2) The limit theorem of the theory of probability does not find any application (at that time they were not quite developed in the theory of probability itself).
(3) The author considers his task not as one of establishing physical theories directly, but as one of constructing statistic-mechanical models which have some analogies in thermodynamics and some other parts of physics; hence he does not hesitate to introduce some very special hypotheses of a statistical character (canonical distribution, ch. 25, § 25) without attempting to prove them or even to interpret their meaning and significance.
(4) The mathematical level of the book is not high; although the arguments are clear from the logical standpoint, they do not pretend to any analytical rigor.

At the time of publication of Gibbs' book, the fundamental problems raised in mathematical science in connection with the foundation of statistical mechanics became more or less clear. If we disregard some isolated small problems, we have here two fundamental groups of problems representing a broad, deep, interesting and difficult field of research in mathematics which is far from being exhausted even at present. The first of these groups is centered around the so-called ergodic problem (ch. III), that is, the problem of the logical foundation for the interpretation of physical quantities by averages of their corresponding functions, averages taken over the phase-space or a suitably selected part of it. This problem, originated by Boltzmann, apparently is far from its complete solution even at the present time. This group of problems was neglected by the investigators for a long time after some unsuccessful attempts, based either on some inappropriate hypotheses introduced ad hoc, or on erroneous logical and mathematical arguments (which, unfortunately, have been repeated without any criticism in later handbooks). In the book of Gibbs these problems naturally are not considered because of the tendency to construct models. Only recently (1931), the remarkable work of G. D. Birkhoff again attracted the attention of many investigators to these problems, and since then this group of problems has never ceased to interest mathematicians, who devote more and more effort to it every year. We will discuss this group of problems in more detail in the ch. III.

The second group of problems is connected with the methods of computation of the phase-averages. In the majority of cases, these averages cannot be calculated precisely. The formulas which are derived for them in the general theory (that is, without specification of the mechanical system under discussion) are complicated, not easy to survey, and as a rule, not suited for mathematical treatment. It is quite natural, therefore, to try to find simpler and more convenient approximations for these averages. This problem is always formulated as a problem of deriving asymptotic formulas which approach the precise formulas when the number of particles constituting the given system increases beyond any limit. These asymptotic formulas have been found long ago by a semi-heuristic method (by means of an unproved extrapolation, starting from some of the simplest examples) and were without rigorous mathematical justification until fairly recent years. A decided change in this direction was brought about by the papers of Darwin and Fowler about twenty years ago. Strictly speaking these authors were the first to give a systematic computation of the average values; up to that time, such a computation was in most cases replaced by a more or less convincing determination of "most probable" values which (without rigorous justification) were assumed to be approximately equal to the corresponding average values. Darwin and Fowler also created a simple, convenient, and mathematically rigorous apparatus for the computation of asymptotic formulas. The only defect of their theory lies in an extreme abstruseness of the justification of their mathematical method. To a considerable extent this abstruseness was due to the fact that the authors did not use the limit theorems of the theory of probability (sufficiently developed by that time), but created anew the necessary analytical apparatus. In any case, the course in statistical mechanics published by Fowler[3] on the basis of this method, represents up to now the only book on the subject, which is on a satisfactory mathematical level.[4]

In closing this brief sketch we should mention that the development of atomic mechanics during the last decades has changed the face of physical statistics to such a degree that, naturally, statistical mechanics had to extend its mathematical apparatus in order to include also quantum phenomena. Moreover, from the modern point of view, we should consider quantized systems as a general type of which the classical systems are a limiting case. Fowler's course is arranged according to precisely this point of view: the new method of constructing asymptotic formulas for phase-averages is established and developed for the quantized systems, and the formulas which correspond to the classical systems are obtained from these by a limiting process.

Quantum statistics also presents some new mathematical problems. Thus, the justification of the peculiar principles of statistical calculations which are the basis of the statistics of Bose-Einstein and Fermi-Dirac required mathematical arguments which were distinct as a matter of principle (not only by their mathematical apparatus) from all those dealt with in the classical statistical mechanics. Nevertheless, it could be stated that the transition from the classical systems to the quantum systems did not introduce any essentially new mathematical difficulties. Any method of justification of the statistical mechanics of the classical systems, would require for quantized

---

[3]Fowler, "Statistical mechanics," Cambridge, 1929.

[4]Except, however, the well known course in the theory of probabilities by v. Mises. However, the main viewpoint of v. Mises differs from the traditional standpoint to such an extent that the theory expounded by him hardly could be given the historically established name of statistical mechanics; mechanical concepts are almost completely eliminated from this theory. In any case, we shall have no occasion to compare the exposition of v. Mises with other expositions.

systems an extension of the analytical apparatus only, in some cases introducing small difficulties of a technical character but not presenting new mathematical problems. In places where we might have to use finite sums or series, we operate with integrals, continuous distributions of probability might be replaced by the discrete ones, for which completely analogous limit theorems hold true.

Precisely for these reasons in the present book we have restricted ourselves to the discussion of the classical systems, leaving completely out of consideration everything concerning quantum physics, although all the methods which we develop after suitable modifications could be applied without any difficulties to the quantum systems. We have chosen the classical systems mainly because our book is designed, in the first place, for a mathematical reader, who cannot always be assumed to have a sufficient knowledge of the foundations of quantum mechanics. On the other hand, we did not consider as expedient the inclusion in the book of a brief exposition of these foundations. Such an inclusion would have considerably increased the size of the book, and would not attain the desired purpose since quantum mechanics with its novel ideas, often contradicting the classical representations, could not be substantially assimilated by studying such a brief exposition.

2. Methodological characterization. Statistical mechanics has for its purpose the construction of a special physical theory which should represent a theoretical basis for some parts of physics (in the first place, for thermodynamics) using as few special hypotheses as possible. More precisely, statistical mechanics considers every kind of matter as a certain mechanical system and tries to deduce the general physical (in particular, thermodynamical) laws governing the behavior of this matter from the most general properties of mechanical systems, and eo ipso to eliminate from the corresponding parts of physics any theoretically unjustified postulation of their fundamental laws. The basic assumptions of statistical mechanics should be then (1) any general laws which hold for all (or at least for very general classes of) mechanical systems, and (2) representations of any kind of matter as a mechanical system consisting of a very large number of components (particles). Thus the purpose of statistical mechanics consists in deriving special properties of such many-molecular systems from the general laws of mechanics and in showing that, with a suitable physical interpretation of the most important quantities appearing in the theory, these derived special properties will give precisely those fundamental physical (and in particular, thermodynamical) laws governing matter in general and certain special kinds of matter in particular. The mathematical method which allows us to realize these aims, for the reasons explained in §1, is the method of the theory of probabilities.

Let us make some further remarks concerning the above described purpose of statistical mechanics.

1. The fact that statistical mechanics considers every kind of matter as a mechanical system and tries to derive all its properties from the general laws of mechanics, often leads to a criticism of being a priori mechanistic. In fact, however, all reproaches of such kind are based on a misunderstanding. Those general laws of mechanics which are used in statistical mechanics are necessary for any motions of material particles, no matter what are the forces causing such motions. It is a complete abstraction from the nature of these forces, that gives to statistical mechanics its specific features and contributes to its deductions all the necessary flexibility. This is best illustrated by the obvious fact that if we modify our point of view on the nature of the particles of a certain kind of matter and on the character of their interaction, the properties of this kind of matter established by methods of statistical mechanics remain unchanged by these modifications because no special assumption was made in the process of deduction of these properties.

The circumstance of being governed by the general laws of mechanics does not lend any specific features to the systems studied in statistical mechanics; as it has been said already, these laws govern any motion of matter, whether it has any relation to statistical mechanics, or not. The specific character of the systems studied in statistical mechanics consists mainly in the enormous number of degrees of freedom which these systems possess. Methodologically, this means that the standpoint of statistical mechanics is determined not by the mechanical nature, but by the particle structure of matter. It almost seems as if the purpose of statistical mechanics is to observe how far reaching are the deductions made on the basis of the atomic structure of matter, irrespective of the nature of these atoms and the laws of their interaction.

2. Since the mechanical basis of statistical mechanics is restricted only by those general laws which hold for any systems (or at least for very general classes of systems), of considerable interest for us (even before the assumption of a large number of components) are the results of the so-called general dynamics, a branch of mechanics whose purpose is the deduction of such laws which hold for all mechanical systems and can be derived from the general laws of mechanics alone. This theory, evidently of a considerable philosophical interest, is of comparatively recent origin. In the past it was usually assumed that the deductions which could be made from the general laws of mechanics were not sufficiently concrete to have any scientific interest. It developed later that the situation was different, and at present the constructions of general dynamics are attracting interest of more and more investigators. In particular, all the above mentioned investigations of Birkhoff and of the increasing number of his disciples belong to this theory. It is particularly interesting to us that the methods (and partially, problems) of general dynamics, even before any assumptions are made concerning the number of degrees of freedom of a system under investigation, show a definitely expressed statistical tendency. This fact is well-known to anyone who has studied investigations in this field with any amount of attention. Thus the fundamental theorem of Birkhoff is formally equivalent to a certain theorem of the theory of probability; conversely, the theory of stationary stochastic

processes, which represents one of the most interesting chapters of the modern theory of probability, formally coincides with one of the parts of the general dynamics.

The reason for this can be easily recognized. The most important problem of general dynamics is the investigation of the dependence of the character of the motion of an arbitrary mechanical system on the initial data, or more precisely the determination of such characteristics of the motion which in one sense or another "almost do not depend" on these initial data. Such a quantity for a great majority of trajectories assumes values very near to a certain constant number. But the expression "for a great majority of trajectories" has the meaning that the set of trajectories which do not satisfy this requirement is metrically negligible in some metric, that is, has for its measure either zero or a very small positive number.

In this sense many propositions of general dynamics are of a peculiarly typical form. They state that for most general classes of mechanical systems the motion is subjected to certain definite conditions, if not for all initial data then at least for a metrically great majority of them. It is known, however, that propositions which can be formulated in such form, in most cases turn out to be equivalent to some theorems of the theory of probability. This theory from a formal point of view could be considered as a group of special problems of the theory of measure, namely such problems as most often deal with the establishment of a metrically negligible smallness of certain sets. It suffices to remember that the majority of propositions of the theory of functions of a real variable concerned with the notions of convergence "in measure", "almost everywhere" etc., finds an adequate expression in the terminology of the theory of probability. Thus it can be stated that even general dynamics which represents the mechanical basis of statistical mechanics, is a science which is filled to a great extent with the ideas of the theory of probability and which successfully uses its methods and analogies.

As to the statistical mechanics, it is a science whose probabilistic character is noticeable in two entirely distinct and completely independent features: in the general dynamics as its mechanical basis, and in the postulate of a great number of degrees of freedom allowing a most fruitful application of methods of the theory of probability.

3. It remains to discuss the form in which methods and results of the theory of probability could be utilized in determining asymptotic formulas which express approximately the phase averages of various functions in the case of a large number of degrees of freedom (or for systems consisting of a large number of particles).

As previously mentioned, in most expositions these formulas are introduced without any justification. After having derived these formulas for some especially simple particular case (for instance, for a homogeneous mono-atomic ideal gas) the authors usually extend them to the general case either without any justification, or using some arguments of heuristical character. Perhaps a single exception from this general rule is represented by the method of Fowler. Darwin and Fowler, as was already mentioned, develop a special and very abstruse analytical apparatus for a mathematical justification of the method of obtaining asymptotic formulas, which they have created. Nowhere do they use explicit results of the theory of probability; instead, they build a separate logical structure, but, as a matter of fact, they are merely moving along an analytical path parallel to that which is used by the theory of probability in deriving its limit theorems. From here only one step remains in attempting to introduce a method which we consider as the most expedient: instead of repeating in a complicated formulation the whole long analytical process which leads to limit theorems of the theory of probability, we attempt to find immediately the bridge which unifies these two groups of problems, and the transition formula which would reduce the entire asymptotic problem of the statistical mechanics to the known limit theorem of the theory of probabilities. This is the path we will take in the present book. In this manner we will be able to achieve simultaneously two ends: from the methodological point of view we will make clear the role of the theory of probabilities in the statistical mechanics; from the formal point of view we have the possibility of establishing the propositions of statistical mechanics on the basis of the mathematically exact laws of the theory of probabilities. In order to emphasize the two above mentioned points we will give in the subsequent text the formulation of the necessary limit theorems of the theory of probabilities without giving their proofs (the latter will be given in the appendix). We hope that such a method of presentation will be attractive to many of those readers who are frightened by the complicated formalistics of the Darwin-Fowler method.

CHAPTER II

GEOMETRY AND KINEMATICS OF THE PHASE SPACE

3. The phase space of a mechanical system. In the statistical mechanics it is convenient to describe the state of a mechanical system G with s degrees of freedom, by values of the Hamiltonian variables $q_1, q_2, \cdots, q_s; p_1, p_2, \cdots, p_s$. The equations of motion of the system then assume the "canonical" form

$$\frac{dq_i}{dt} = \frac{\partial H}{\partial p_i}, \qquad \frac{dp_i}{dt} = -\frac{\partial H}{\partial q_i}, \qquad (1 \leq i \leq s), \tag{1}$$

where H is the so-called Hamiltonian function of the $2s$ variables $q_1, \cdots, p_s$ (we always shall assume it not to depend on time explicitly). The function $H(q_i, p_k)$ is an integral of the system (1). Indeed, in view of this system of equations,

$$\frac{dH}{dt} = \sum_{i=1}^{s} \frac{\partial H}{\partial q_i}\frac{dq_i}{dt} + \sum_{i=1}^{s} \frac{\partial H}{\partial p_i}\frac{dp_i}{dt}$$

$$= \sum_{i=1}^{s} \frac{\partial H}{\partial q_i}\frac{\partial H}{\partial p_i} - \sum_{i=1}^{s} \frac{\partial H}{\partial p_i}\frac{\partial H}{\partial q_i} = 0.$$

Since system (1) contains only equations of the first order, the values of the Hamiltonian variables $q_1, \cdots, p_s$ given for some time $t = t_0$, determine their values for any other time t (succeeding or preceding t_0).

Imagine now a Euclidean space Γ of $2s$ dimensions, whose points are determined by the Descartes coordinates $q_1, \cdots, p_s$. Then to each possible state of our mechanical system G there will correspond a uniquely determined point of the space Γ, which we shall call the image point of the given system; the whole space Γ we agree to call the phase space of this system. We shall see that, for the purposes of the statistical mechanics, the geometrical interpretation of the set of all possible states of the system by means of its space, appears exceedingly fruitful and receives a basic methodological significance.

Since the state of the system at any given time determines uniquely its state at any other time, the motion of the image point in the phase space which represents the changes of state of the given system depending on time is uniquely determined by its initial position. The image point describes in the phase space a curve which we shall call a trajectory. It follows that through each point of the phase space there passes one and only one trajectory, and the kinematic law of motion of the image point along this trajectory is uniquely determined.

If at the time t_0 the image point of the system G is some point M_0 of the space Γ, and at the other time t (succeeding or preceding t_0) some other point M, then the points M_0 and M determine each other uniquely. We can say that the point M_0 of the phase space during the interval of time $(t_0\,,\,t)$ goes over into M. During the same interval of time every other point of the space Γ goes over into a definite new position, in other words all this space is transformed into itself and in one-to-one way, since, conversely, the position of a point at the time t determines uniquely its position at the time t_0. Furthermore if we keep t_0 fixed and vary t arbitrarily, we see that all the set of possible changes of state of the given system is represented as a continuous sequence (one-parameter group) of transformations of its phase space into itself, which sequence can be considered as a continuous motion of this space in itself. This representation also turns out to be very convenient for the purposes of the statistical mechanics. We shall call the above-described motion of the phase space in itself its natural motion. Since the displacement of any point of the phase space in its natural motion during an interval of time Δt, depends only on the initial position of this point and the length of this interval, but does not depend on the choice of the initial time, the natural motion of the phase space is stationary. This means that the velocities of points of the phase space in this motion depend uniquely on the position of these points, but do not change with the time.

In what follows, we shall often call the Hamiltonian variables $q_1, \cdots, p_s$ of the given system G the dynamic coordinates of its image point in the space Γ, and any function of these variables the phase function of the given system. The most important phase function is the Hamiltonian function $H(q_1, \cdots, p_s)$. This function determines completely the mechanical nature of the given system, because it determines completely the system of equations of motion. In particular, this function determines completely the natural motion of the phase space of the given system.

When convenient we shall denote the set of the dynamic coordinates of the given mechanical system (the point of the phase space) by a single letter P, and, correspondingly, an arbitrary phase function by $f(P)$.

There are cases where the phase space Γ has a part Γ' with the property that an arbitrary point of this part remains in it during all the natural motion of the space Γ. Such a part Γ' participates in the natural motion by transforming into itself, and therefore it is called an invariant part of the space Γ. In what follows we shall see that the motion of an invariant part plays a very essential role in the statistical mechanics.

The special form of the Hamiltonian system (1) has as a consequence the fact, easy to foresee, that not every continuous transformation of the phase space into itself can appear as its natural motion. A natural motion is characterized by some special properties, and the most important of these properties can be formulated in two theorems on which, to a considerable extent, is based the whole construction of the statistical mechanics. We shall pass now to a proof of these theorems.

4. Theorem of Liouville. The first of these two theorems (under slightly more restricted assumptions) was proved by the French mathematician Liouville in the middle of the past century.

Let M be any measurable (in the sense of Lebesgue) set of

points of the phase space Γ of the given mechanical system. In the natural motion of this space the set M goes over into another set M_t during an interval of time t. The theorem of Liouville asserts that the measure of the set M_t for any t coincides with the measure of the set M. In other words, the measure of measurable point-sets is an invariant of the natural motion of the space Γ.

For the proof of this theorem it will be convenient to introduce a more uniform notation for the dynamic coordinates of points of the space Γ. Let

$$x_i = q_i, \qquad x_{s+i} = p_i \qquad (i = 1, 2, \cdots, s)$$

and

$$X_i = \frac{\partial H}{\partial p_i}, \qquad X_{s+i} = -\frac{\partial H}{\partial q_i} \qquad (i = 1, 2, \cdots, s).$$

In this notation the canonical system (1) of the §3 can be written in the form

$$\frac{dx_i}{dt} = X_i(x_1, x_2, \cdots, x_{2s}) \qquad (1 \le i \le 2s). \tag{2}$$

For what follows let us observe that

$$\sum_{i=1}^{2s} \frac{\partial X_i}{\partial x_i} = \sum_{i=1}^{s} \frac{\partial^2 H}{\partial p_i\, \partial q_i} - \sum_{i=1}^{s} \frac{\partial^2 H}{\partial q_i\, \partial p_i} = 0. \tag{3}$$

If $x_i^{(0)}$ $(i = 1, 2, \cdots, 2s)$ are the values of the variable, x_i at some definite time t_0, we obtain as a uniquely determined solution of the system (2), the system of functions

$$x_i = f_i(t; x_1^{(0)}, \cdots, x_{2s}^{(0)}) \qquad (1 \le i \le 2s).$$

Let us agree to denote the measure of the set A by $\mathfrak{M}A$. Then

$$\mathfrak{M}M_t = \int_{M_t} dx_1 \cdots dx_{2s}.$$

In this integral let us change the variables by setting

$$x_i = f_i(t; y_1, \cdots, y_{2s}),$$

where t is considered as an auxiliary parameter. Since the point $(y_1, \cdots, y_{2s})$ of the space Γ obviously describes the set M when the point $(x_1, \cdots, x_{2s})$ describes the set M_t,

$$\mathfrak{M}M_t = \int_M J(t; y_1, \cdots, y_{2s})\, dy_1 \cdots dy_{2s},$$ [1]

where

$$J = J(t; y_1, \cdots, y_{2s}) = \frac{\partial(x_1, \cdots, x_{2s})}{\partial(y_1, \cdots, y_{2s})}.$$

If we differentiate this with respect to t we find

$$\text{(4)} \qquad \frac{d\mathfrak{M}M_t}{dt} = \int_M \frac{\partial J}{\partial t}\, dy_1 \cdots dy_{2s}.$$

We can compute $\partial J/\partial t$ by the rule of differentiation of determinants. We find

$$\text{(5)} \qquad \frac{\partial J}{\partial t} = \sum_{i=1}^{2s} J_i,$$

where

$$J_i = \frac{\partial(x_1, \cdots, x_{i-1}, \partial x_i/\partial t, x_{i+1}, \cdots, x_{2s})}{\partial(y_1, \cdots, y_{2s})} \quad (1 \le i \le 2s).$$

In view of the system (2), since $\partial x_i/\partial t$ coincides with dx_i/dt,

$$J_i = \frac{\partial(x_1, \cdots, x_{i-1}, X_i, x_{i+1}, \cdots, x_{2s})}{\partial(y_1, y_2, \cdots, y_{2s})} \quad (1 \le i \le 2s).$$

But

$$\frac{\partial X_i}{\partial y_k} = \sum_{r=1}^{2s} \frac{\partial X_i}{\partial x_r}\frac{\partial x_r}{\partial y_k} \quad (1 \le i \le 2s,\ 1 \le k \le 2s),$$

---

[1] In this book we shall, in most cases, denote multiple integrals using only one integral sign; the dimensionality of the integral will be determined either by the number of differentials under the integral sign, or by some other obvious considerations. In cases where the domain of integration is not indicated explicitly, the integration will be taken over the whole space.

hence the previous equality gives

$$J_i = \sum_{r=1}^{2s} \frac{\partial X_i}{\partial x_r} \frac{\partial(x_1, \cdots, x_{i-1}, x_r, x_{i+1}, \cdots, x_{2s})}{\partial(y_1, \cdots, y_{2s})} \qquad (1 \leq i \leq 2s).$$

But clearly

$$\frac{\partial(x_1, \cdots, x_{i-1}, x_r, x_{i+1}, \cdots, x_{2s})}{\partial(y_1, \cdots, y_{2s})} = \begin{cases} \zeta, & \text{if} \quad r = i, \\ 0, & \text{if} \quad r \neq i \end{cases}$$

whence

$$J_i = J \frac{\partial X_i}{\partial x_i} \qquad (1 \leq i \leq 2s).$$

On substituting into (5) and using (3) we have

$$\frac{\partial J}{\partial t} = J \sum_{i=1}^{2s} \frac{\partial X_i}{\partial x_i} = 0.$$

Then (4) shows

$$\frac{d\mathfrak{M}M_t}{dt} = 0$$

which proves the invariance of the measure in the natural motion of the phase space.

Corollary. In the natural motion of the phase space every point P, during an interval of time t, goes over into a uniquely determined other point which we always shall denote by P_t. If $f(P)$ is an arbitrary phase function, we shall write

$$f(P_t) = f(P, t)$$

here t might be also negative. Now, let M be a Lebesgue measurable set of points of the space Γ, of finite measure, and $f(P)$ a phase function, Lebesgue integrable over Γ. By the Liouville theorem the volume element of the space Γ during the time t goes over into an equal volume element dV_t. Let us consider the integral

$$\int_{M_t} f(P)\, dV_t$$

and let us change the variables by introducing as new variables the dynamic coordinates of the point which goes over into P during the time t. It is clear that: (1) the new domain of integration will be the set M; (2) under the sign of integral the symbolic argument P should be replaced by P_t; (3) the element dV_t should be replaced by the equal element dV. Thus we get

$$\int_{M_t} f(P)\, dV_t = \int_M f(P_t)\, dV = \int_M f(P,\, t)\, dV.$$

In the left hand member we also can write dV instead of dV_t, so that

$$\int_{M_t} f(P)\, dV = \int_M f(P,\, t)\, dV. \tag{6}$$

In particular, if the set M is invariant, then, for each t,

$$\int_M f(P)\, dV = \int_M f(P,\, t)\, dV. \tag{7}$$

5. *Theorem of Birkhoff.* The second theorem, to the proof of which we have to turn now, was proved comparatively recently (in 1931) by G. D. Birkhoff (the form of the proof which we are giving here is due to A. N. Kolmogoroff).

Let V be an invariant part of the phase space of a finite volume, $f(P)$ a phase function summable over V and determined at all points[2] $P \in V$.

The theorem of Birkhoff asserts that the limit

$$\lim_{C \to \infty} \frac{1}{C} \int_0^C f(P,\, t)\, dt$$

exists for all points P of the set V, except at most of a certain set of measure zero (or, more concisely, almost everywhere on V). The limit also exists almost everywhere when $C \to -\infty$.

---

[2]The notation $a \in A$ means that a is an element of the set A.

It is clear that we can interpret the quantity $(1/C)\int_0^C f(P, t)\,dt$ as the average of the function $f(P)$ along the trajectory passing through P during the interval of time $(0, C)$. The limit of this expression if $C \to \infty$ we shall call the time average of the function $f(P)$ along the trajectory passing through P. The theorem of Birkhoff asserts that, for a summable function, the time averages exist along the trajectories passing through almost all points of V.

We now pass to the proof. In what follows we set for any integer n

$$\int_n^{n+1} f(P, t)\,dt = x_n = x_n(P),$$

$$\int_n^{n+1} | f(P, t) |\,dt = y_n = y_n(P).$$

Lemma 1. Almost everywhere on V for $n \to \infty$

$$\frac{1}{n}\, y_n(P) \to 0.$$

The proof of this lemma is based on the Liouville theorem. If we introduce a new variable α in the integral defining y_n, determined by $t = n + \alpha$, we get

$$\begin{aligned} y_n(P) &= \int_0^1 | f(P, n + \alpha) |\,d\alpha \\ &= \int_0^1 | f(P_n, \alpha) |\,d\alpha = y_0(P_n) \end{aligned} \tag{8}$$

since obviously $f(P, n + \alpha) = f(P_n, \alpha)$.

Let us denote by $E_{n,n}$ and $E_{n,0}$ the sets of points P belonging to V and satisfying respectively the conditions

$$y_n(P) > \epsilon n \qquad \text{and} \qquad y_0(P) > \epsilon n,$$

where ϵ is an arbitrary fixed positive number. It is easy to see that in the natural motion of the space Γ the set $E_{n,n}$ during the time n goes over into $E_{n,0}$. Indeed, in view of (8) the inequality $y_n(P) > \epsilon n$ is equivalent to $y_0(P_n) > \epsilon n$. Hence $P \in E_{n,n}$ implies $P_n \in E_{n,0}$, and conversely. Therefore, by Liouville's theorem,

$$\mathfrak{M}E_{n,n} = \mathfrak{M}E_{n,0}\,. \tag{9}$$

Now we show that the series

$$\sum_{n=1}^{\infty} \mathfrak{M}E_{n,n}$$

converges. In view of (9) this series can be written as

$$\sum_{n=1}^{\infty} \mathfrak{M}E_{n,0}\,.$$

If we denote by F_m the set of points of V for which $m\epsilon < y_0(P) \leq (m+1)\epsilon$ and observe that $E_{n,0} = \sum_{m=n}^{\infty} F_m$ we can write this series in the form

$$\begin{aligned}
\sum_{n=1}^{\infty}\sum_{m=n}^{\infty} \mathfrak{M}F_m &= \sum_{m=1}^{\infty}\sum_{n=1}^{m} \mathfrak{M}F_m = \sum_{m=1}^{\infty} m\mathfrak{M}F_m \\
&= \frac{1}{\epsilon}\sum_{m=1}^{\infty} m\epsilon\mathfrak{M}F_m \leq \frac{1}{\epsilon}\sum_{m=1}^{\infty}\int_{m\epsilon<y_0\leq(m+1)\epsilon} y_0(P)\,dV \\
&\leq \frac{1}{\epsilon}\int_V y_0(P)\,dV = \frac{1}{\epsilon}\int_V dV\int_0^1 |\,f(P,\alpha)\,|\,d\alpha \\
&= \frac{1}{\epsilon}\int_0^1 d\alpha\int_V |\,f(P,\alpha)\,|\,dV
\end{aligned}$$

where dV is the element of volume of the space Γ.

Since V is an invariant part of the space Γ, the last expression, in view of (7), is

$$\frac{1}{\epsilon}\int_0^1 d\alpha\int_V |\,f(P)\,|\,dV = \frac{1}{\epsilon}\int_V |\,f(P)\,|\,dV.$$

Since $f(P)$ by assumption is summable over V, this is a finite number, which proves our assertion.

By a known theorem of the metric theory of sets it follows that every point of the set V, except at most a set of measure zero, belongs to no more than a finite number of sets of the sequence $E_{n,n}$ $(n = 1, 2, \cdots)$. In other words, for almost all points $P \in V$ there exists a number $N = N(P)$ such that for each $n > N$,

$$y_n(P) \leq \epsilon n.$$

Since ϵ is arbitrary, Lemma 1 is proved.

Let now for any $a < b$,

$$h_{ab}(P) = \frac{1}{b-a} \int_a^b f(P, t)\, dt.$$

By the definition of x_n, if a and b are integers, we have

$$h_{ab}(P) = \frac{1}{b-a} (x_a + x_{a+1} + \cdots + x_{b-1}).$$

Lemma 2. If $h_{0n}(P)$, as $n \to \infty$ assuming integer values, has no limit on a set M of positive measure, then there exist two real numbers α and β ($\alpha < \beta$) and a part M^* of the set M, such that $\mathfrak{M}M^* > 0$ and at each point $P \in M^*$,

$$l(P) = \liminf_{n \to \infty} h_{0n}(P) < \alpha,$$

$$L(P) = \limsup_{n \to \infty} h_{0n}(P) > \beta.$$

This proposition which is almost self-evident, can be easily proved. Let us consider the set of all intervals δ_n (α_n, β_n) with rational end points (the order of numeration is immaterial). If $P \in M$ then $l(P) < L(P)$[3], and therefore among

[3]Except for the cases where $h_{0n}(P) \to +\infty$ or $h_{0n}(P) \to -\infty$. It is easy to see however that for a summable function this can occur only on a set of measure zero. Indeed, if we had, say, $h_{0n}(P) \to \infty$ on a set of positive measure, then, by a known theorem of Egoroff we could assert the uniformity of this process on a certain other set N, $\mathfrak{M}N > 0$. Let $A > 0$ be arbitrarily large and let for $n > n_0 = n_0(A)$, $h_{0n}(P) > A$ on N. Assuming $n > n_0$ and integrating over N, in view of (6) we have

$$A\mathfrak{M}N \leq \int_N h_{0n}(P)\, dV = \frac{1}{n} \int_0^n d\alpha \int_N f(P, \alpha)\, dV$$

$$= \frac{1}{n} \int_0^n d\alpha \int_{N_\alpha} f(P)\, dV \leq \int_V |f(P)|\, dV$$

which leads to a contradiction since A is arbitrarily large.

the intervals δ_n there will be found the first one, say δ_m, for which

$$l(P) < \alpha_m < \beta_m < L(P).$$

Denote by M_m the set of all points P of M which are connected in this sense with the interval δ_m. It is clear that

$$M = \sum_{m=1}^{\infty} M_m$$

and that the sets M_i and M_k for $i \neq k$ have no points in common. Since $\mathfrak{M}M > 0$, we will have $\mathfrak{M}M_m > 0$ for at least one value of m. On setting $\alpha = \alpha_m$, $\beta = \beta_m$, $M_m = M^*$ we see that Lemma 2 is proved.

Assume now that the conditions of Lemma 2 are satisfied. Let $P \in M^*$ and consider an interval (a, b) where $a < b$ are integers. We shall call this segment a proper segment of the point P if the following conditions are satisfied:

(1) $\quad h_{ab}(P) > \beta,$

(2) $\quad h_{ab'}(P) \leq \beta \quad$ for $\quad a < b' < b.$

We shall show that two proper segments (a_1, b_1) and (a_2, b_2) of the same point P cannot partially overlap each other. Indeed, if we had for instance $a_1 < a_2 < b_1 < b_2$, then we would have

$$(b_1 - a_1)h_{a_1b_1} = (a_2 - a_1)h_{a_1a_2} + (b_1 - a_2)h_{a_2b_1}$$

while, by (2),

$$h_{a_1a_2} \leq \beta, \qquad h_{a_2b_1} \leq \beta, \qquad h_{a_1b_1} > \beta$$

which would lead to the contradictory relation

$$\beta(b_1 - a_1) < \beta(a_2 - a_1) + \beta(b_1 - a_2) = \beta(b_1 - a_1).$$

Furthermore, let us agree to call a proper segment of P a maximal proper segment of rank s, if its length does not exceed s, and if it is not contained in any other proper segment of P whose length does not exceed s. It is easy to see that every proper segment of length not exceeding s is contained in one and only one maximal proper segment of rank s. Indeed, among all the proper segments of length not exceeding s and containing the given segment, there will be one of maximal length. It is clear that this will be a maximal proper segment of rank s. Its uniqueness follows from the fact that if there existed two such segments, then they would have points in common (as containing the given segment). But then either one of them would be contained in the other, and therefore would not be a maximal segment of rank s, or they would partially overlap, which has just been proved to be impossible.

For every positive integer s let us denote by M_s the set of points P of the set M^* for which the inequality

$$h_{0n}(P) > \beta$$

holds for at least one $n \leq s$. It is obvious that every $P \in M^*$ belongs to all M_s when s is sufficiently large, so that

$$M^* = \sum_{s=1}^{\infty} M_s$$

and, since $M_s \subset M_{s+1}$,

$$\mathfrak{M}M^* = \lim_{s \to \infty} \mathfrak{M}M_s .$$

But $\mathfrak{M}M^* > 0$, hence, for s sufficiently large, also $\mathfrak{M}M_s > 0$. In what follows we shall denote by s some fixed positive integer satisfying this condition.

Lemma 3. In order that P would belong to the set M_s it is necessary and sufficient that P would have a maximal proper segment (a, b) of rank s, such that $a \leq 0 < b$.

Proof. 1) Let $P \in M_s$ and let n be the smallest positive integer for which $h_{0n}(P) > \beta$, so that $n \leq s$. Then clearly, the segment $(0, n)$ is a proper segment of P. As it has been proved above, this segment is contained in a unique maximal proper segment of rank s, which satisfies all conditions of Lemma 3.

2) Let P have a maximal proper segment (a, b) of rank s, such that $a \leq 0 < b$. To complete the proof of Lemma 3 it is

sufficient to show that in this case $h_{0b}(P) > \beta$, because $b \leq b - a \leq s$.

If $a = 0$, our statement is obvious because (a, b) is a proper segment of the point P, whence $h_{ab}(P) > \beta$. If $a < 0$,

$$(b - a)h_{ab}(P) = (x_a + \cdots + x_1) + (x_0 + \cdots + x_{b-1}),$$

or

$$(b - a)h_{ab}(P) = -ah_{a0}(P) + bh_{0b}(P),$$

whence

$$h_{0b}(P) = \frac{1}{b}[(b - a)h_{ab}(P) + ah_{a0}(P)].$$

But, by definition of a proper segment,

$$h_{ab}(P) > \beta, \qquad h_{a0}(P) \leq \beta,$$

and since $b - a > 0$ and $a < 0$,

$$h_{0b}(P) > \frac{1}{b}[(b - a)\beta + a\beta] = \beta.$$

Consider now any point P of the set M_s and a maximal proper segment (a, b) of rank s corresponding to P in the sense of Lemma 3, so that $a \leq 0 < b$, $b - a \leq s$. Let $b - a = q$, $-a = p$, then $1 \leq q \leq s$, $0 \leq p \leq q - 1$. In what follows we denote by δ_{pq} the segment $(-p, -p + q)$ and by M_{pq} the set of all points of M_s which correspond to the segment δ_{pq} in the sense of Lemma 3, so that

$$M_s = \sum_{q=1}^{s} \sum_{p=0}^{q-1} M_{pq} .$$

It is easy to see that in the natural motion of the space Γ the set M_{0q} after p units of time goes over into M_{pq} (because $h_{0q}(P) = h_{-p,q-p}(P_p)$). Thus

$$\mathfrak{M}M_{pq} = \mathfrak{M}M_{0q} \qquad (0 \leq p \leq q - 1).$$

It is also clear that the sets M_{pq} with different pairs of indices cannot have points in common. Finally, in view of formula (6), the same relationship between the sets M_{0q} and M_{pq} shows that for any summable function $\varphi(P)$,

$$\int_{M_{pq}} \varphi(P)\, dV = \int_{M_{0q}} \varphi(P, p)\, dV.$$

From all that has been said above we conclude

$$\begin{aligned}
\int_{M_s} x_0(P)\, dV &= \sum_{q=1}^{s} \sum_{p=0}^{q-1} \int_{M_{pq}} x_0(P)\, dV \\
&= \sum_{q=1}^{s} \sum_{p=0}^{q-1} \int_{M_{0q}} x_0(P_p)\, dV \\
&= \sum_{q=1}^{s} \sum_{p=0}^{q-1} \int_{M_{0q}} dV \int_0^1 f(P_p\,,\, t)\, dt \\
&= \sum_{q=1}^{s} \sum_{p=0}^{q-1} \int_{M_{0q}} dV \int_p^{p+1} f(P, \alpha)\, d\alpha \\
&= \sum_{q=1}^{s} \int_{M_{0q}} dV \int_0^q f(P, \alpha)\, d\alpha \\
&= \sum_{q=1}^{s} \int_{M_{0q}} qh_{0q}(P)\, dV > \beta \sum_{q=1}^{s} q\mathfrak{M}M_{0q} \\
&= \beta \sum_{q=1}^{s} \sum_{p=0}^{q-1} \mathfrak{M}M_{pq} = \beta\mathfrak{M}M_s
\end{aligned}$$

This relation holds for all s sufficiently large; on allowing $s \to \infty$ we get;

$$\int_{M^*} x_0(P)\, dV \geq \beta\mathfrak{M}M^*. \tag{10}$$

Since for all points of M^* we also have

$$\liminf_{n\to\infty} h_{0n}(P) < \alpha,$$

we can prove in the same way that

$$\int_{M^*} x_0(P)\, dV \leq \alpha\mathfrak{M}M^*. \tag{11}$$

Since $\alpha < \beta$, the inequalities (10) and (11) are contradictory, which shows that our assumption $\mathfrak{M}M^* > 0$ is not possible, or in other words, that the limit

$$\lim_{n\to\infty} h_{0n}(P)$$

must exist almost everywhere.

To accomplish the proof of Virkoff's theorem it remains to remove the restriction that the parameter n assumes only integral values. This is easily done by means of Lemma 1. Indeed, since the expression

$$\frac{1}{b}\int_0^{[b]} f(P, t)\, dt$$

(where $[b]$ is the largest integer contained in b), as $b \to \infty$ differs from $(1/[b]) \int_0^{[b]} f(P, t)\, dt$ only by a factor which tends to 1, and since the latter expression has a limit almost everywhere, the limit

$$\lim_{b\to\infty} \frac{1}{b}\int_0^{[b]} f(P, t)\, dt$$

also exists almost everywhere. On the other hand

$$\left| \frac{1}{b}\int_0^{b} f(P, t)\, dt - \frac{1}{b}\int_0^{[b]} f(P, t)\, dt \right|$$

$$\leq \frac{1}{b}\int_{[b]}^{b} |\, f(P, t)\, |\, dt \leq \frac{1}{[b]}\int_{[b]}^{[b]+1} |\, f(P, t)\, |\, dt$$

$$= \frac{y_{[b]}(P)}{[b]} \to 0 \qquad (b \to \infty)$$

by Lemma 1. Hence the limit

$$\lim_{b\to\infty} \frac{1}{b}\int_0^{b} f(P, t)\, dt$$

also exists almost everywhere, which completes the proof of the theorem of Birkhoff.

6. Case of metric indecomposability. We shall call the quantity

$$\hat{f}(P) = \lim_{C\to\infty} \frac{1}{C} \int_0^C f(P, t)\, dt$$

the "time average" of the function $f(P)$ along the trajectory passing through P. Such a terminology, strictly speaking, becomes suitable only after we know that this quantity does not depend on the choice of the initial point on the given trajectory, in other words, that for all P and t

$$\hat{f}(P_t) = \hat{f}(P)$$

(assuming of course that $\hat{f}(P)$ exists). We shall prove this property.

Let, for definiteness, $t > 0$. Since, by assumption, the limit

$$\lim_{C\to\infty} \frac{1}{C+t} \int_0^{C+t} f(P, \alpha)\, d\alpha = \hat{f}(P)$$

exists, and since the difference

$$\frac{1}{C} \int_0^{C+t} f(P, \alpha)\, d\alpha - \frac{1}{C+t} \int_0^{C+t} f(P, \alpha)\, d\alpha$$

$$= \frac{t}{C} \frac{1}{C+t} \int_0^{C+t} f(P, \alpha)\, d\alpha \to 0$$

we also have

$$\lim_{C\to\infty} \frac{1}{C} \int_0^{C+t} f(P, \alpha)\, d\alpha = \hat{f}(P). \tag{12}$$

But

$$\frac{1}{C} \int_0^C f(P_t, \alpha)\, d\alpha = \frac{1}{C} \int_0^C f(P, t+\alpha)\, d\alpha$$

$$= \frac{1}{C} \int_t^{t+C} f(P, \alpha)\, d\alpha = \frac{1}{C} \int_0^{t+C} f(P, \alpha)\, d\alpha - \frac{1}{C} \int_0^t f(P, \alpha)\, d\alpha.$$

In the right-hand member the first term tends to $\hat{f}(P)$, by (12), and the second term tends to 0 as $C \to \infty$. Hence

$$\lim_{C \to \infty} \frac{1}{C} \int_0^C f(P_t, \alpha)\, d\alpha = \hat{f}(P).$$

By definition this limit is $\hat{f}(P_t)$, which proves our assertion.

We turn now to the discussion of the most important special case of Birkhoff's theorem. Let V be some invariant part (of finite volume) of the space Γ. We shall call this part metrically indecomposable if it cannot be represented in the form

$$V = V_1 + V_2$$

where V_1 and V_2 are invariant parts of positive measure. In order to understand clearly the content of this notion, observe that the set V, as any invariant set, is a certain set of complete trajectories. If by any method we separate this set of trajectories into two other sets (each consisting again of complete trajectories), then, if V is metrically indecomposable, only one of the following two cases is possible: either one of the component parts has measure zero (hence the other has measure $\mathfrak{M}V$), or both components are not measurable. In the case the set V is metrically indecomposable, Birkhoff's theorem can be made considerably more precise.

Theorem: If the set V is metrically indecomposable, then almost everywhere on V

$$\hat{f}(P) = \frac{1}{\mathfrak{M}V} \int_V f(P)\, dV.$$

The quantity in the right-hand member of this equality could be interpreted as an average of the function f on the set V. We shall call it the phase average of the function f (on the set V), and denote it by $\bar{f}$. Thus the above stated theorem asserts that, in the case of the metric indecomposability of the set V, the time average $\hat{f}(P)$ of any summable function f, for almost all initial points P is the same, and coincides with the phase average $\bar{f}$ of the same function.

In order to prove our theorem, let us prove first that the

$\hat{f}(P)$ is constant almost everywhere on V. Otherwise, there would exist such a real number α that in splitting V into two parts V_1 and V_2 which are defined respectively by the conditions $\hat{f}(P) > \alpha$ on V_1 and $\hat{f}(P) \leq \alpha$ on V_2, we would have $\mathfrak{M}V_1 > 0$ and $\mathfrak{M}V_2 > 0$.[4] But, by what was proved at the beginning of this paragraph, the sets V_1 and V_2 are invariant, which implies a contradiction to the set V being metrically indecomposable. Thus $\hat{f}(P)$ almost everywhere on V has the same constant value which we denote by a. It remains to prove that $a = \bar{f}$.

$$f_C(P) = \frac{1}{C} \int_0^C f(P, t)\, dt$$

We have

$$a = \frac{1}{\mathfrak{M}V} \int_V a\, dV$$

$$= \frac{1}{\mathfrak{M}V} \int_V [a - f_C(P)]\, dV + \frac{1}{\mathfrak{M}V} \int_V f_C(P)\, dV.$$

By the invariance of the set V (see (7)),

$$\frac{1}{\mathfrak{M}V} \int_V f_C(P)\, dV = \frac{1}{C\mathfrak{M}V} \int_0^C dt \int_V f(P, t)\, dV$$

$$= \frac{1}{C\mathfrak{M}V} \int_0^C dt \int_V f(P)\, dV = \frac{1}{\mathfrak{M}V} \int_V f(P)\, dV = \bar{f}.$$

---

[4]In order to prove this, for any positive integer n, let us subdivide the axis of reals into segments $(k/2^n, (k+1)/2^n)(-\infty < k < \infty)$, and let us call such a segment an essential segment, if the set of points $P \in V$ for which values of $\hat{f}(P)$ belong to this segment, has a positive measure. If for some value of n there exist two essential segments, our assertion is proved. If, however, for every n there exists only one essential segment δ_n, then clearly $\delta_{n+1} < \delta_n$, so that the sequence of segments $\delta_n(n = 1, 2, \cdots)$ has a single point in common α. It is quite obvious that, in this case, $\hat{f}(P) = \alpha$ almost everywhere on V.

Hence

$$a = \frac{1}{\mathfrak{M}V} \int_V [a - f_C(P)]\, dV + \bar{f}$$

and the quantity

$$\frac{1}{\mathfrak{M}V} \int_V [a - f_C(P)]\, dV = a - \bar{f}$$

does not depend on C. Our theorem will be proved if we show that this integral is equal to zero.

Let $\epsilon > 0$ be arbitrarily small. Let $V_1(C)$ be the set of those points $P \in V$ for which

$$| a - f_C(P) | < \epsilon$$

while

$$V_2(C) = V - V_1(C).$$

It is clear that

$$\left| \int_V [a - f_C(P)]\, dV \right| \leq \int_{V_1(C)} | a - f_C(P) |\, dV$$

(13)
$$+ \int_{V_2(C)} | a - f_C(P) |\, dV$$

$$\leq \epsilon \mathfrak{M}V + | a | \mathfrak{M}V_2(C) + \int_{V_2(C)} | f_C(P) |\, dV.$$

Since $f_C(P)$ tends to a almost everywhere on V as $C \to \infty$, $\mathfrak{M}V_2(C) \to 0$ as $C \to \infty$ (convergence in measure). Thus, for C sufficiently large, we have $\mathfrak{M}V_2(C) < \epsilon$. But

$$\int_{V_2(C)} | f_C(P) |\, dV \leq \frac{1}{C} \int_0^C dt \int_{V_2(C)} | f(P, t) |\, dV$$

(14)
$$= \frac{1}{C} \int_0^C dt \int_{V_2(C,t)} | f(P) |\, dV$$

where $V_2(C, t)$ is the set into which $V_2(C)$ goes over during the interval of time t, in the natural motion of the phase space. By the theorem of Liouville, for all t,

$$\mathfrak{M}V_2(C, t) = \mathfrak{M}V_2(C),$$

and $\mathfrak{M}V_2(C, t) \to 0$ as $C \to \infty$, uniformly with respect to t. In view of the absolute continuity of integrals of summable functions, we can take C so large that, for all t,

$$\int_{V_2(C,t)} | f(P) | \, dV < \epsilon$$

Then (14) also shows that, in this case,

$$\int_{V_2(C)} | f_C(P) | \, dV < \epsilon$$

and then (13) gives

$$\left| \int_V [a - f_C(P)] \, dV \right| \leq \epsilon \mathfrak{M}V + | a | \epsilon + \epsilon.$$

Thus the left-hand member of this inequality will be as small as we like if C is sufficiently large, and since it does not depend on C, it must be equal to zero, which had to be proved.

7. *Structure functions.* From point of view of physics, the most important phase function of the given mechanical system is its total energy

$$E = E(q_1, \cdots, q_s; p_1, \cdots, p_s).$$

For an isolated system this function has a constant value, in other words, represents an integral of the system of equations of motion. Therefore, for any constant a the region of the phase space for the point of which $E = a$, is an invariant part of the phase space. For simplicity, we shall call such regions surfaces of constant energy. We shall consider only such cases where the function E has a lower bound over the whole space Γ (this is the case for the most interesting physical systems).

Using the arbitrariness of choice of the addition constant in the expression of the potential energy (which enters as a term in the expression of the function E), we may assume that a lower bound of E is zero, so that $E \geq 0$ on the whole space Γ. Furthermore we always shall assume that the portion of the phase space characterized by the inequality $E < x$, for each $x > 0$ is a simply connected domain bounded by the surface $E = x$. This surface we shall assume to be closed and sufficiently smooth to justify analytical methods which will be applied in various problems. We shall denote by Σ_x the surface of constant energy $E = x$. For $x_1 < x_2$ the surface Σ_{x_1} is situated entirely inside Σ_{x_2}, so that, schematically, the family of surfaces of constant energy could be represented as a family of concentric hyper spheres. In the natural motion of the phase space each surface of constant energy, and also each domain bounded by two such surfaces, is transformed (into itself), in other words, is an invariant part of the phase space.

All assumptions made above are satisfied by the systems which are usually considered in the physical applications of the statistical mechanics. Moreover, the total energy of such systems coincides with the Hamiltonian function. It follows first that if E is given as a function of the dynamic variables, the mechanical nature of the given system is completely determined. Secondly, we can then use the argument which we used in §3 to prove that the Hamiltonian function is an integral of the system of equations of motion, as a proof of the law of conservation of energy for the systems we are going to consider.

Let us denote by $V(x)$ the volume of the part V_x of the space Γ, in which $E < x$ (that is, of the domain inside the surface Σ_x). $V(x)$ is a monotone function which increases from 0 to ∞ as x varies between the same limits. If $x_1 < x_2$ the volume of the layer enclosed between the surfaces Σ_{x_1} and Σ_{x_2}, is equal to $V(x_2) - V(x_1)$.

In what follows we shall use the following theorem.

Theorem: Let $f(P)$ be a point function in the space Γ, summable over a certain domain contained inside the domain V_x.

Then

$$\frac{d}{dx}\int_{V_x} f(P)\, dV = \int_{\Sigma_x} f(P)\, \frac{d\Sigma}{\text{grad } E}.$$

(Here dV and $d\Sigma$ denote respectively the volume elements of the space Γ and of the surface Σ_x, and

$$\text{grad } E = \left\{ \sum_{k=1}^{s} \left[\left(\frac{\partial E}{\partial q_k}\right)^2 + \left(\frac{\partial E}{\partial p_k}\right)^2 \right] \right\}^{1/2}$$

is the gradient of E.)

For the proof observe that the element of the volume dV of the domain V_x in the left-hand member can be replaced by the product $dn\, d\Sigma$, where $d\Sigma$ is the volume element of that surface of constant energy to which abuts the element dV, and dn the element of the outward normal to this surface (the thickness of the layer separating this surface from another surface immediately adjoining). Such a change signifies merely a special choice of the subdivision of the domain V_x, which we use to construct the integral, the choice which is characterised by the fact that initially the domain is subdivided in infinitely thin layers by a net of surfaces of constant energy. Thus

$$\int_{V_x} f(P)\, dV = \int_{V_x} f(P)\, d\Sigma\, dn.$$

To simplify, let us denote the dynamic coordinates of the point P by $x_1, x_2, \cdots, x_{2s}$, as it was done in §4. Since

$$dx_i = dn \cos(n, x_i) \qquad (1 \le i \le 2s),$$

the increment of the energy when we pass from some surface of constant energy to an infinitely near surface will be given by the formula

$$dE = \sum_{i=1}^{2s} \frac{\partial E}{\partial x_i}\, dx_i = dn \sum_{i=1}^{2s} \frac{\partial E}{\partial x_i} \cos(n, x_i).$$

But it is known that

$$\cos(n, x_i) = \frac{\partial E/\partial x_i}{\text{grad } E} \qquad (1 \le i \le 2s)$$

whence

$$dE = dn \frac{\sum_{i=1}^{2s} \left(\frac{\partial E}{\partial x_i}\right)^2}{\text{grad } E} = dn \text{ grad } E$$

and we get

$$\int_{V_x} f(P)\, dV = \int_{V_x} f(P) \frac{d\Sigma\, dE}{\text{grad } E} = \int_0^x dE \int_{\Sigma_E} f(P) \frac{d\Sigma}{\text{grad } E}.$$

The value of the inner integral here is a function of E. On denoting this function by $\psi(E)$ we get

$$\frac{d}{dx} \int_{V_x} f(P)\, dV = \frac{d}{dx} \int_0^x \psi(E)\, dE = \psi(x) = \int_{\Sigma_x} f(P) \frac{d\Sigma}{\text{grad } E}$$

as was to be proved.

Since, by the law of conservation of energy, each surface Σ_x of constant energy of the space Γ, is an invariant part of the space Γ, in the natural motion of this space, every measurable set M situated on this surface, during any interval of time goes over into another measurable set situated on the same surface. However, if we define the measure of the set M by

$$\text{(15)} \qquad \mathfrak{M}M = \int_M d\Sigma$$

this measure, in general, would not remain invariant. The set M moving on the surface Σ_x in the natural motion of the phase space, would at the same time change its measure. Such a situation would have been extremely inconvenient for our theory, since, in discussing motion on the surface Σ_x we would have been deprived of such valuable tools as the theorem of Liouville, Birkhoff and their corollaries. That is why in the statistical mechanics the definition (15) of measure on the surface Σ_x is always replaced by another definition which is invariant with respect to the natural motion of the space Γ. After such a replacement we can consider each surface of constant energy as a bounded region, invariant to the natural motion of which all the results obtained in preceding paragraphs can be applied. In the construction of our theory which follows we make precisely this choice.

In order to obtain an invariant definition of measure on the given surface Σ_x of constant energy $E = x$, consider any measurable (in the sense of (15)) set M on it. At each point of this set draw the outward normal to the surface Σ_x to its intersection with the infinitely near surface $\Sigma_{x+\Delta x}$. The part of the space Γ which is filled by these normals is bounded and will be denoted by D. The volume

$$\int_D dV$$

of this part is clearly invariant with respect to the natural motion of the phase space and can be represented also in the form

$$\int_{x<E<x+\Delta x} f(P)\, dV,$$

where $f(P)$ has value 1 or 0 according as P does, or does not, belong to D. The ratio of this volume to Δx and also the limit of this ratio as $\Delta x \to 0$ are also invariants of the natural motion of the space Γ. But by the theorem which just has been proved this limit is

$$\int_{\Sigma_x} f(P) \frac{d\Sigma}{\text{grad } E} = \int_M \frac{d\Sigma}{\text{grad } E}.$$

Thus if we define

$$\mathfrak{M}M = \int_M \frac{d\Sigma}{\text{grad } E}$$

we will have an invariant definition of measure on the surface Σ_x. This definition of measure we shall use in all that follows (it is obvious that it satisfies all conditions which a definition of measure has to satisfy).

In particular, the measure (volume) $\Omega(x)$ of the whole surface Σ_x will be

$$\Omega(x) = \int_{\Sigma_x} \frac{d\Sigma}{\text{grad } E}. \tag{16}$$

If we put $f(P) = 1$ in the general theorem proved above, we obtain

$$\Omega(x) = V'(x). \tag{17}$$

Thus, the measure of the whole surface of constant energy, with our definition of measure, is simply equal to the derivative with respect to x of the volume of the domain V_x of phase space bounded by this surface. This fact considerably simplifies the geometry of the structure of the phase space, in which we are interested now.

According to our definition of measure we shall interpret the expression

$$\bar{f} = \frac{1}{\Omega(x)} \int_{\Sigma_x} f(P) \frac{d\Sigma}{\text{grad } E}$$

as the average of any summable function $f(P)$ defined on Σ_x. This is the limit, as $\Delta x \to 0$, of the average of $f(P)$ on the layer enclosed between the surfaces Σ_x and $\Sigma_{x+\Delta x}$. By the theorem proved above this average can be also represented in the form

$$\bar{f} = \frac{1}{\Omega(x)} \frac{d}{dx} \int_{V_x} f(P)\, dV.$$

This formula in many cases turns out very convenient for evaluation of averages of phase functions on surfaces of constant energy.

The function $\Omega(x)$ defined by (16) is a monotone function increasing from 0 to ∞[5] as x varies between the same limits. As we shall see later, this function completely determines the most important features of the mechanical structure of the corresponding physical system. In what follows we shall call this function the structure function of the given system. Therefore, the structure function of the given system can be

[5] *Footnote of the translator.* This appears as an additional assumption.

defined either as the measure of the surface of constant energy (with our special definition of measure) or as the derivative with respect to x of the function $V(x)$ defined above.

8. Components of mechanical systems. In this paragraph, as we have done several times before, we denote by x_1, x_2, $\cdots$, x_{2s} the dynamical coordinates of a point of the space Γ, where the order of numeration is irrelevant. Each phase function and, in particular, the total energy E of the given system, is a function of these $2s$ variables.

Suppose that the function

$$E = E(x_1, \cdots, x_{2s})$$

can be represented as a sum of two terms E_1 and E_2 where the first term depends on some (not all) of the dynamical coordinates and the second term depends on the remaining coordinates. Since the order of numeration of the dynamical coordinates is irrelevant we may write $E = E_1 + E_2$ where

$$E_1 = E_1(x_1, x_2, \cdots, x_k),$$

$$E_2 = E_2(x_{k+1}, x_{k+2}, \cdots, x_{2s}).$$

In such a case we agree to say that the set $(x_1, x_2, \cdots, x_{2s})$ of the dynamical coordinates of the given system is decomposed in two components $(x_1, \cdots, x_k)$ and $(x_{k+1}, \cdots, x_{2s})$. We could express it also by saying that the given system consists of two "components" which appear as bearers of the corresponding sets of the dynamic coordinates. From the point of view of the formal theory it does not make any difference whether we call a component of the given system the set of coordinates $(x_1, \cdots, x_k)$ itself, or if we attribute to this set a certain "bearer" to which we shall give the name of a component. We shall use both terminologies without danger of any confusion. From a more realistic point of view it appears natural to try to interpret each component as a separate physical system which is contained in the given system. However, such a viewpoint will be too narrow, and in some cases will not suit our purposes. The point is that, although each materially isolated part of our system determines in most cases a certain component of this system, it is useful to consider occasionally such components (that is sets of coordinates) to which there does not correspond any materially isolated part of the system. The isolated character of such components is of a purely energy nature, the precise sense being given by the above definition of a component. For instance, if the system consists of one material particle the components of the velocity of which and the mass are respectively u, v, w, m, and if its energy E reduces to the kinetic energy

$$E = \frac{m}{2}(u^2 + v^2 + w^2),$$

we could consider the quantity u as a component of our system, and formally attribute to it a certain "bearer" whose energy is $(mu^2)/2$, although in this case there is no question of any material bearer (we shall see later that such considerations can prove to be useful).

In any case, if to each component of the given system we may attribute a definite energy, (from the definition of a component), then each component, being essentially a group of dynamic coordinates, has its own phase space, and the state of this component (that is the set of values of its coordinates) is represented by a point of this phase space. The phase space Γ of the given system is clearly the direct product of the phase spaces Γ_1 and Γ_2 of its two components, and the volume element of the space can be taken to be equal to the product $dV_1\, dV_2$ of the volume elements of the spaces Γ_1 and Γ_2.

Furthermore, each component has its own structure function. The law of composition of structure functions, that is the formula which determines the structure function of the given system in terms of structure functions of its components, is one of the most important basic formulas of our theory. We now pass on to the derivation of this formula.

First we make the following observation: If we have a phase function of the given system whose value is completely determined by the value of the energy of the system at the corresponding point of the space Γ, then the integral of such a function $f(E)$, taken over the domain of the space Γ enclosed between two surfaces of constant energy Σ_{x_1} and Σ_{x_2}, can be easily expressed in the form of a simple integral, namely,

$$\int_{x_1<E<x_2} f(E)\,dV = \int_{x_1}^{x_2} f(x)\,\Omega(x)\,dx. \tag{18}$$

Indeed, we may evaluate our multiple integral by subdividing the domain $x_1 < E < x_2$ of the space into infinitely thin layers between surfaces of constant energy. In the layer between the two surfaces Σ_x and $\Sigma_{x+\Delta x}$ the function $f(E)$ (which for simplicity is supposed to be continuous) can (with an infinitesimal error) be assumed to be equal to $f(x)$, while the volume of this layer, up to infinitesimals of higher order, is

$$V'(x)\Delta x = \Omega(x)\Delta x$$

which gives formula (18). In particular,

$$\int_{\Gamma} f(E)\,dV = \int f(x)\,\Omega(x)\,dx. \tag{19}$$

This formula is used in a great number of applications.

Now let $V(x)$ and $\Omega(x)$ be the functions as defined above for the given system, while $V_1(x)$, $\Omega_1(x)$ and $V_2(x)$, $\Omega_2(x)$ the corresponding functions for the two components of the given system. Then

$$V(x) = \int_{V_x} dV = \int_{(V_x)_1} dV_1 \int_{(V_x - E_1)_2} dV_2$$

$$= \int_{(V_x)_1} V_2(x - E_1)\,dV_1,$$

where $(V_x)_1$ denotes the set of all points of the space Γ, at which $E_1 < x$, and $V_{(x-E_1)_2}$ is defined in an analogous fashion.

Since the phase function $V_2(x - E_1)$ of the first component depends only on its energy E_1, by formula (18) we have

$$V(x) = \int_0^x V_2(x - E_1)\,\Omega_1(E_1)\,dE_1$$

and since $V_2(x - E_1) = 0$ for $E_1 > x$, the integration can be extended to infinity so that

$$V(x) = \int_0^\infty V_2(x - y)\Omega_1(y)\, dy.$$

Finally, on differentiating this with respect to x, we have

$$\Omega(x) = \int_0^\infty \Omega_1(y)\Omega_2(x - y)\, dy. \tag{20}$$

This is the law of composition of structure functions which we intended to establish.

All that has been said above, without any modifications, can be extended to the case where the given system consists not of two, but of any number of components. The definition of component remains unchanged. As before, the space Γ is the direct product of the phase spaces of all the components. For the law of composition of structure functions, in case of n components, we have the formula

$$\Omega(x) = \int \left\{ \prod_{i=1}^{n-1} \Omega_i(x_i)\, dx_i \right\} \Omega_n\left(x - \sum_{i=1}^{n-1} x_i\right), \tag{21}$$

where the integration is extended over the whole space of $(n - 1)$ dimensions (or over the domain $x_i > 0, 1 \leq i \leq n - 1$, which is the same, since $\Omega_i(x_i) = 0$ for $x_i < 0$). To derive this formula it is simplest to use the method of complete induction from n to $(n + 1)$, by decomposing the n-th component in (21) into two components, and by expressing the last factor in terms of structure functions of these two components, using formula (20).

To conclude these brief preliminary considerations, we remark that the conception of decomposition of the system into components, leads to a specific methodological paradox, as has been observed several times. As stated already at the beginning of Chapter I, with all the generality and abstractness of the hypotheses of the statistical mechanics, it is invariably assumed that particles of the matter are in a state of intensive energy interaction, where the energy of one particle is transferred to another (for instance by means of collisions). As we shall see in more detail later, the statistical mechanics bases its method precisely on a possibility of such an exchange of energy between various particles constituting the matter. However, if we take the particles constituting the given physical system to be the components in the above defined sense, we are excluding the possibility of any energetical interaction between them. Indeed, if the Hamiltonian function, which expresses the energy of our system, is a sum of functions each depending only on the dynamic coordinates of a single particle (and representing the Hamiltonian function of this particle), then, clearly, the whole system of equations (1) splits into component systems each of which describes the motion of some separate particle and is not connected in any way with other particles. Hence the energy of each particle, which is expressed by its Hamiltonian function, appears as an integral of equations of motion, and therefore remains constant.

The serious difficulty so created is resolved by the fact that we can consider particles of matter as only approximately isolated energetically components. There is no doubt that a precise expression for the energy of the system must contain also terms which depend simultaneously on the energy of several particles (mutual potentials of particles), and which assure the possibility of an energetical interaction between the particles (from a mathematical point of view, prevent the splitting of the system (1) into systems referring to single particles). But, inasmuch as forces of interaction between the particles manifest themselves only at very small distances, such mixed terms in the expression of energy, representing mutual potential energy of particles, will be (in the great majority of points of the phase space) negligible as compared with the kinetic energy of particles or with the potential energy of external fields. In particular they will contribute very little in evaluating various averages, hence in the majority of computations in statistical mechanics we will be able to neglect such terms, and, to a good approximation, assume that the energy of the system is equal to the sum of the energies of constituent particles, these thus appearing as components of our system in the above defined sense. However, these mixed terms which are neglected, from the point of principle play a very important role, since it is precisely their presence that assures the possibility of an exchange of energy between the particles, on which is based the whole method of the statistical mechanics.

CHAPTER III

ERGODIC PROBLEM

9. Interpretation of physical quantities in statistical mechanics. The values of physical quantities which characterize the state of the system we are studying are determined uniquely by this state, which, in turn, is described in our theory by the set of the dynamic coordinates. Thus a physical quantity, as a rule, appears as a function of the dynamic coordinates of the system, or, what amounts to the same thing, a function of a point of its phase space as its phase function. Therefore, if, we wish to compare the deductions of our theory with the experimental data from measurements of various physical quantities, we will compare the values of various physical quantities found experimentally with the values of the corresponding phase functions furnished by our theory. However, such a statement of the problem leads immediately to a series of methodological difficulties which threaten to leave this problem without any content. The point is that the phase functions of the system in general represent quantities which assume widely distinct values for different states of the system. In order to compare these values with experimental data we should have a possibility of determining the state of the system at the time of the experimental measurement, that is, to determine the values of all the dynamic coordinates for this time. For instance, in the case of a gas, this would mean to determine at least the positions and the velocities of all constituent molecules, a problem which obviously is insoluble. If we forsake this idea, then what states of the system we should assume in order to compute those values of the phase functions which will have to be compared with the experimental data is an entirely open question.

The following considerations will allow us to alleviate to a certain extent the acuteness of this difficulty. An experiment or an observation which gives the measurement of a physical quantity is performed not instantaneously, but requires a certain interval of time which, no matter how small it appears to us, would, as a rule, be very large from the point of view of an observer who watches the evolution of our physical system. This system will be subjected during this interval of time to various perturbations (such as mutual collisions of molecules) which may change essentially the values of the corresponding phase function. Thus we will have to compare experimental data not with separate values of phase functions, but with their averages taken over very large intervals of time. In other words, according to what was said in the preceding chapter, with time averages of phase functions over a trajectory which represents the evolution of our physical system.

This consideration of course changes the picture quite considerably, but, at the same time, immediately introduces new difficulties. The first of these arises from the fact that the time averages of a given phase function taken over a given trajectory may have widely distinct values for different time intervals. This difficulty is alleviated considerably by the theorem of Birkhoff, which states that, for almost all trajectories, the time averages of the given phase function, which tend to a definite limit when the time interval tends to infinity, will assume approximately the same value for all time intervals, sufficiently large. It is therefore natural to take this limit as the time average furnished by our theory.

There is, however, another difficulty which is much harder to overcome, namely, that we cannot determine which trajectory in the phase space is traversed by our system. If this system has s degrees of freedom (where s, as a rule, is a very large number), in order to determine this trajectory we would need to find values of $(2s - 1)$ integrals of the system, which do not depend on time, while actually we can determine approximately values of only very few of these integrals. (The value of the energy of the system almost always is considered to be given.) The determination of any integral gives us in the phase space a surface which contained the trajectory in mention. If we know the values of k of such integrals, then we know that our trajectory belongs to a certain "reduced" manifold of $(2s - k)$ dimensions, so that for $k = 2s - 1$ the trajectory will be completely determined; if however, as usually happens, we know only one integral of energy, then $k = 1$, and the only thing we can say about the trajectory, is that it belongs to a manifold of $(2s - 1)$ dimensions (surface of constant energy).

There is however a case where this difficulty does not exist, in view of the theorem of section 6. If the given surface at the constant energy is metrically indecomposable, then the time averages of any summable function are the same for almost all trajectories and coincide with the phase average of this function on the given surface of constant energy. In this case every physical quantity receives a definite interpretation in our theory as the phase average of the corresponding phase function, and the above mentioned difficulties no longer exist.

Actually, in all expositions of the statistical mechanics, this phase average is taken as a theoretical interpretation of any physical quantity. In doing so either no arguments at all are given in favor of such a choice, or a special hypothesis is constructed in order to justify this choice, or, finally, various reasons are cited in favor of such an interpretation, indicating at the same time that these reasons are not logically obligatory and that the interpretation was generally accepted in view of the successful results to which the theory based on this interpretation leads. The last method appears to us most preferable scientifically, and, in the following paragraphs of this chapter, we shall attempt to discuss in detail the most important questions pertaining to the subject, from the point of view of modern ideas.

At present we remark in addition that, in view of what has been said above, the task of a mathematical justification of the statistical mechanics reduces essentially to two problems. The first of these two problems, is to investigate as exhaustively as possible, under what conditions and to what degree the time averages of phase function, which, as we have seen, appear as a natural interpretation of experimental measurements, can be replaced by the phase averages of the same functions. The desirability, and even inevitability, of such a replacement is clear: The computation of time averages requires the knowledge of trajectories, that is, the complete integration of the equations of motion and determination of all the constants of integration, which of course cannot be done for systems considered in statistical mechanics, with their large numbers of degrees of freedom. As said before, we shall discuss the question connected with this first problem in the following paragraphs of the present chapter.

The second problem which will be considered in the next chapters, is to create a general method for approximate computation of phase averages or surfaces of constant energy. The evaluation of phase averages, contrary to the evaluation of time averages, is a problem completely accessible to a mathematical analysis, although it involves certain difficulties. This problem is always formulated as a problem of constructing a general method which would allow us to derive sufficiently simple asymptotic formulas for phase averages, under the assumption that the number of degrees of freedom of the given system increases beyond limit. Since statistical mechanics deals with systems with very large degrees of freedom, we may expect that such asymptotic expressions will be sufficiently near to precise values of phase averages.

10. Fixed and free integrals. The problem of a theoretical justification of the replacement of time averages by phase averages, is usually called the *ergodic problem* (sometime this terminology is used for other related problems.) Almost always, one considers the averages of phase functions on a given surface of constant energy. Therefore, in attempting to give a short account of the history and of the present status of the ergodic problem we first should try to understand why in our theory we choose precisely these phase averages. From a purely theoretical point of view, such a choice at first glance, appears casual and arbitrary. Usually such a concept of phase averages is justified by the following argument. Since the energy is an integral of the equations of motion, each trajectory is situated entirely on some surface of constant energy, Σ_x. The values of the function under consideration, at the points of the phase space Γ which are not on this surface Σ_x, play no role in evaluating the time averages, and therefore should not be taken into account in evaluating the phase averages, if we desire that these phase averages be near the time averages.

Such an argument contains a vulnerable point. Everything which is said therein about the energy integral can be repeated, word for word, for any other integral of motion, which does not depend on time. Since, for a system with s degrees of freedom, there are $(2s - 1)$ of such independent integrals, we should fix the values of each of them beforehand, or, in other words, determine the trajectory of the system in the phase space, and evaluate our averages along this trajectory. This, however, is never done, and is not feasible to do, because the great majority of other integrals of motion are not known, so that we cannot determine the trajectory which represents the evolution of our system.

Thus the whole question requires more careful consideration. It will be most convenient for us to start by making more precise the above argument in favor of a preliminary specification of a surface of constant energy Σ_x. In itself this argument is not only entirely convincing, but serves as a starting point of our discussion.

Suppose that we do not specify the surface Σ_x, but try to evaluate the phase averages of our functions over the whole space Γ. The first, and comparatively non-essential difficulty here is due to the fact that this space has an infinite volume, so that averages of simplest functions would become infinite or undetermined, if we would not introduce a preliminary weighting of the space with the purpose of diminishing the contributions by distant portions of the space. However, such a weighting of various parts of the space Γ would necessarily introduce some element of arbitrariness, which would make the computation of phase averages based on this weighting, somewhat doubtful.[1]

However, this difficulty, as we have observed before, is not essential in comparison with the other difficulty which apparently makes the whole method completely useless. Indeed, the energy of the given system is a phase function, and undoubtedly, one of the most important ones. Our method should attribute to it some definite average value $\overline{E}$, as for any other phase function. But what physical meaning could this average have? In particular, could we expect that the time average of the energy of the given system will be equal to $\overline{E}$ (or at least near to $\overline{E}$) in the majority of the evolutionary processes of which the system is capable? It is sufficient to formulate this assumption to understand its absurdity. In each evolutionary process the given isolated system preserves a constant value of its energy. This constant value we can select, in general, arbitrarily over a rather wide range, and in different cases we can select different values which are quite far from any fixed number prescribed by the theory. The very attempt to attribute to the energy of our system any fixed value, no matter by what method this value is computed, contradicts reality. And so, the preliminary reduction of the phase space to some surface at constant energy appears really inevitable for any efficient evaluation of the phase averages.

Let us now investigate why we may pay no attention to all other integrals independent of time and treated in the same way as we have treated the energy. Such claims, as we have already observed before, appear well founded, at least at first sight. However, a more attentive consideration will show that

---

[1]Here the question is one of introducing weights "universal" for the given system. A weighting adjusted to a definite value of the energy (or of some other integral) as is done, for instance, in the so-called canonical distribution of Gibbs (see Chapter V, section 25) is equivalent to a preliminary specification of the surface Σ_x, and therefore is not of interest to us at present.

the situation is different. For better understanding we shall split our argument into several steps.

1. As we shall see later, the majority of physical phase functions with which we have to deal in statistical physics have a specific structure which makes the values of such a function, defined on every surface Σ_x, very near each other at all points, except for a set of a very small measure. This implies that for the majority of the trajectories situated on Σ_x, the time averages of such a function will have values very near each other, and therefore near the phase average of the function over the surface Σ_x.

2. Let now I be any integral of the equations of motion of the given system which does not depend on time and is distinct from the energy integral. If, considered as a phase function it has a structure described under 1, then the possibility of replacing the time average by the phase average for this integral does exist. If, however, I does not have such a structure, then the phase function which it represents, as a rule, will not have actual physical interpretation, and therefore the relationship between its various averages will not present any interest for us.

3. It is possible however that in some cases the arguments of 2 cannot satisfy us even if they are quite correct. Such cases occur when the integral I represents a physical quantity which plays a role analogous to the role of the energy, that is a quantity for which we are able to select a value arbitrary within certain limits, by regulating the conditions of our process, or at least are able to determine its value experimentally. For instance, let $\bar{I}$ be the phase average of the function I over the surface Σ_x. In view of what has been said in 1 and 2, the time averages of the function I for most trajectories will be near $\bar{I}$. But if, for various reasons, we forced I to assume a value I_0 which is far from $\bar{I}$, or if we have found such a value by experimental measurement, then of course we have to attribute to I the value I_0, not $\bar{I}$. The fact that *in most cases* I is near $\bar{I}$, cannot force us to accept this relation if we know that *in our case* (whether due to our interference or not) the

value I is far from $\bar{I}$. Furthermore, if we know the actual value of I, we will have to take it into account in computing values of other phase functions. In other words, in such a case the value of the integral I could and should be specified beforehand, just as has been done with the integral of energy.[2]

To abbreviate, let us call the integrals of type described above—*controllable* integrals (because we can either select, or determine experimentally, its value; in other words we control its value in the process we consider). Let our system have k such controllable integrals (they will amost always include the integral of energy). If we fix the value of each of them in our process, we shall specify in the phase space of our system a certain *reduced manifold* of $2s - k$ dimensions, over which we will have to take the phase averages of the phase functions in which we are interested. In the great majority of cases dealt with in the statistical physics, the only controllable integral is the energy integral, so that the reduced manifold will be only the surfaces of constant energy Σ_x. There are, however, cases where, simultaneously with the energy integrals, some other integrals become controllable (for instance integrals of the momentum components). In such cases the phase averages are actually taken over the manifolds of smaller number of dimensions, which are obtained by fixing the values of controllable integrals.

As concerns the remaining free (that is, not fixed) integrals, each of them, if it represents an actual physical quantity, will be almost constant, in the above described sense, on the reduced manifold, as stated in 2. This gives us a certain reason to expect that its value will be near its phase average on the reduced manifold, in the majority of cases met in practice. In other words, we assume that the location of the image point of a system on the reduced manifold is a random event such that a very small probability corresponds to the location of the

---

[2]An excellent example of how radically all the results of computations could be changed by fixing the value of such an integral, is given by the statistical schemes of Bose-Einstein and Fermi-Dirac in quantum physics.

point in a set of very small measure (absolute continuity!). Hence it is almost certain that our integral assumes values near to its average, in the majority of experimental measurements. Of course, the question of correctness of all this hypothetical construction can be ultimately decided only by a comparison of the deductions of our theory with the experimental data.

The fact that the distinction between the fixed and free integrals is determined not by their mathematical nature, but, so to say, by their role in our scientific or practical experience, should not at all disturb a mathematician. It is typical in all applications of the theory of probability. For instance, under normal conditions we consider the number of tickets drawn in a lottery as a random variable. However, if we succeed in studying the mechanism of the drawing to such extent that we shall be able to determine this number beforehand, or, still more, if we succeed in drawing the number as we desire, then all elements of randomness disappear, although the mechanism of drawing is the same in both cases.

In what follows we shall consider as a reduced manifold of the phase space the surface of constant energy, which corresponds to the actual situation for the majority of systems discussed in the statistical physics. Thus it will be our purpose to collect as many arguments as possible in favor of the fact that the time averages of physically most important phase functions, for the great majority of trajectories situated on the given surface of constant energy have values which are close to each other (and therefore, necessarily, near the values of the corresponding phase averages).

11. Brief historical sketch. As we have indicated already, many authors attempted to prove the coincidence of the time and the phase averages by introducing various special hypotheses, more or less plausible. Such hypotheses usually were called "ergodic hypotheses". The first of them was stated by Boltzmann, who also was the first to use the terminology. Boltzmann, conjectured that each surface of constant energy consists of a single trajectory. In other words, no matter what is the state of our system at a given time, it will pass (or has already passed) through any other state with the same value of the total energy.

Using this conjecture it is possible to establish the coincidence between the time and phase averages on each surface of constant energy. However, the conjecture itself is logically contradictory, which soon was found out, and which at present is topologically obvious, since no trajectory can have multiple points and therefore cannot fill out the whole many dimensional space.

After this failure attempts were made for a long time to replace the ergodic hypothesis of Boltzmann by the "quasiergodic" hypothesis, according to which every trajectory, although not filling completely the surface of constant energy on which it is situated, constitutes an everywhere densepoint set (that is intersects every element of the surface). However, even if we disregard the fact that the logical compatibility of this hypothesis has not been established, nobody succeeded in proving on this basis the possibility of replacement of the time averages by the phase averages. Numerous expositions of such proofs contain grave mistakes. Those authors (as, for instance, P. Kertz in his known treatise) who do not wish to base their proofs on false arguments, have to introduce several additional hypotheses.

All this history of the ergodic problem appears to us instructive since it makes the efficacy of introducing various hypotheses which are not supported by any argument very doubtful. As is usual in such cases, when we are not able to submit really convincing arguments in favor of replacing the time averages by the phase averages, it is preferable, and also simpler, to attempt as the "ergodic hypothesis" the very possibility of such a replacement, and then to judge the theory constructed on the basis of this hypothesis, by its practical success or failure. This, of course, does not mean that the theoretical justification of the accepted hypothesis is to be forgotten. On the contrary, this question remains one of the most fundamental in the statistical mechanics. We wish only to say that the reduction of this hypothesis to others is little justified, and does not appear to us to be very efficient.

After several decades of almost fruitless discussions in connection with the ergodic problem, it was only in 1931 that the theorem of Birkhoff revived the problem.* From our point of view the essence of these new researches consists in emphasizing the notion of the metrical indecomposability and in establishing its close connection with the ergodic problem. Quite often one hears, particularly from the side of physicists, that Birkhoff's results do not give anything for the solution of the ergodic problem, but only reduce it to another problem—the justification of the metric indecomposability of the surfaces of constant energy, and, in this sense, are similar to introducing ergodic hypotheses as done by earlier authors. Although we do not wish to overestimate the role of Birkhoff's results for the foundation of the statistical mechanics, we cannot share such a point of view.** The enormous, interesting, and significant literature which has developed on the basis of Birkhoff's researches during the last decade shows that these researches in the ergodic problem shed light on new problems which remained unknown beforehand, and discovered a most fertile field for new researches. In all mathematical justifications of various special fields usually there occur moments when, although it does not solve any concrete problem, the introduction of some appropriate notions coordinates and organizes the whole problem in such a fundamental way that the work of an investigator is turned from a chaotic and almost helpless wandering into a sensible and planned conquest of new scientific facts. There is every reason to believe that the researches of

---

**Footnote of the translator*. The first form of an ergodic theorem was a somewhat weaker statement proved by J. v. Neumann shortly before Birkhoff.

***Footnote of the translator*. In fact, a general theorem showing the *existence* of ergodic transformations on quite general manifolds (or phase spaces) was proved by I. Oxtoby and S. Ulam. (Ann. of Math. 42, 874 (1941)). Their results imply that in a certain sense *almost every* continuous transformation is metrically transitive.

Birkhoff represent such a moment in the development of the ergodic problem. In the following paragraphs we intend to state some observations concerning the present status of the ergodic problem.

12. On metric indecomposability of reduced manifolds. As we know from section 6 of chapter II the metric indecomposability of the given surface of constant energy insures that the time averages of any summable function $f(P)$ along almost all trajectories situated on this surface coincide with the phase average $\bar{f}$ of this function over this surface. But it is easy to see that if we require this property of almost all trajectories and all summable functions, then the condition of metric indecomposability is also necessary. Indeed, if the given surface of constant energy is metrically decomposable, then it can be split in two invariant sets M_1 and M_2, each of which has a positive measure. The summable function $\varphi(P)$ which assumes the value 0 on M_1, and the value 1 on M_2 cannot have the same time averages along almost all trajectories (these time averages are either 0 or 1, while the phase average has some intermediary value).

Thus we see that the metric indecomposability of the surfaces of constant energy is a necessary and sufficient condition for a positive answer to the ergodic problem stated in a certain precisely determined sense. This fact alone shows the essential advantage of the investigations of Birkhoff as compared with the introduction of old "ergodic hypotheses".

We now pass on to the question of what general considerations may make the metric indecomposability of surfaces of constant energy more or less plausible. Let $\varphi = \varphi(q_1, \cdots, p_s)$ one of the "free" integrals of the equations of motion of the given system, that is, an integral independent of the integral of energy and not containing time explicitly. If the function φ remained constant on each surface of constant energy, then the value of the integral φ would have been uniquely determined by the value of the integral of energy, and these two integrals would not be independent, contrary to our assumption. Hence the function φ cannot remain constant on each surface of constant energy, and, being continuous, cannot remain constant at almost all points of such a surface.

Let us take a surface of constant energy on which the function φ is not constant almost everywhere. Then we can find such a real number α that each of the two parts of this surface, characterized respectively by the inequalities $\varphi > \alpha$ and $\varphi \leq \alpha$, will be of positive measure.[3] But since φ is an integral of the equations of motion of our system, each of these two parts is an invariant set. This shows that our surface of constant energy cannot be metrically indecomposable.

This elementary argument leaves an impression that the metric indecomposability of surfaces of constant energy is a hypothesis, which, like the Boltzmann hypothesis, never can be realized, and therefore should be rejected. However, this would mean a complete solution of the ergodic problem in the negative sense.

Formally, there are no objections against the above argument, and we actually have to agree that the answer to the ergodic problem at least in the form in which it was formulated above, should be in the negative. We shall see however, that if we introduce some sensible and natural modifications in the formulation of the problem, we may obtain a positive answer, at least in some cases.

So far it was always self-evident, although not stated explicitly, that two distinct points of the phase space represent two distinct states of our mechanical system. Actually, however, in many cases, to distinct points of the space Γ may correspond identical states of the mechanical system.

Let us explain this. In many cases we are forced to characterize the same physical state of the system not by one, but by several sets (sometimes even by infinitely many) of values of its dynamic coordinates. Thus for a point which moves uniformly along a circumference, if we determine its position by the central angle counted from some fixed radius, we must

[3]For the proof see the footnote on p. 30.

consider as identical the states for which the values of this angle differ by a multiple of 2π.

On the other hand it is obvious that every physical quantity which characterizes the state of the given system must be determined uniquely by this state. The phase function which interprets this physical quantity in our theory, must therefore assume the same value at any two points of the phase space corresponding to the same state of the given system. We shall call *normal* every phase function which satisfies this condition.

Since, in view of the preceding considerations, all physical quantities for which there may arise the question of comparison of their theoretical values with experimental data are interpreted by normal phase functions, we will lose nothing if in formulating the ergodic problem we shall state that not all but only normal summable functions should satisfy its requirement. Then the condition of metric indecomposability will cease to be necessary; it will be replaced by a broader necessary and sufficient condition which can be easily formulated.

We shall call *normal* every subdivision of the given surface of constant energy in two invariant paths of positive measure, such that all points of the surface which correspond to the same state of the system (we shall call such points physically equivalent) belong to the same part of the surface. A surface which does not admit of a normal subdivision we shall call *metrically indecomposable in the extended sense.*

Theorem. In order that time averages of any normal summable function taken along almost all trajectories situated on the given surface of constant energy, would coincide with the phase average of this function over the given surface, it is necessary and sufficient that this surface is metrically indecomposable in the extended sense.

A proof of this theorem can be carried through in complete analogy with the arguments of section 6 of the preceding chapter (sufficiency) and of the beginning of the present paragraph (necessity). We leave it to the reader.

After this modification the ergodic problem is reduced to the question of whether, generally speaking, the surfaces of constant energy of the mechanical systems under consideration are metrically indecomposable in the extended sense. First we shall show that the argument which was used above in establishing the impossibility of the metric indecomposability in the original sense, does not give anything directly if we assume the metric indecomposability in the extended sense.

Indeed, in this argument we subdivided the given surface of constant energy into two parts, placing a point in the one or the other of these parts according to the value assumed at this point by a certain integral φ. But if now we are interested only in the normal subdivisions associated with a given point all physically equivalent points must belong to the same part. If φ is a normal integral, that is, assumes the same value at all physically equivalent points, our argument remains valid. But if our integral φ is not normal, then, in determining the sets M_1 and M_2 we cannot start by arbitrarily subdividing the set of all values assumed by the integral φ in two parts. If we want the subdivision (M_1, M_2) of the surface Σ_x to be normal, we must see to it that the values assumed by φ at any two physically equivalent points are always placed in the same part. This requirement (as we shall see in an example) may turn out to be incompatible with the requirement that M_1 and M_2 be invariant sets of positive measure. In such a case our argument becomes invalid, and the question of possibility of metric indecomposability in the extended sense remains open.

Later we shall give the simplest of known examples of such a situation. Now we note that the above argument shows the impossibility of metric indecomposability even in extended sense, if among the free integrals there exists at least one normal integral. In particular, the energy integral being always normal, necessarily has to be fixed. If the system has no other normal integrals of motion, then we can raise the question of the metric indecomposability in the extended sense on the surfaces of constant energy.

Let us turn now to an example of the metric indecomposability in the extended sense. Consider a system with two degrees of freedom, whose situation is determined by two cyclic coordinates φ, ψ, with a period 1. This means that for any integers k and l, the pair of coordinates $\varphi + k$, $\psi + l$ represents the same situation of the system as the pair φ, ψ (motion of a particle on the surface of a focus). The Hamiltonian function we take to be

$$H = (1/2)(\varphi'^2 + \psi'^2),$$

where φ', ψ' are dynamic coordinates canonically conjugate to φ, ψ. If we denote by a dot the differentiation with respect to time, we can write the canonical system of the equations of motion in the form

$$\dot{\varphi} = \varphi', \qquad \dot{\psi} = \psi', \qquad \dot{\varphi}' = 0, \qquad \dot{\psi}' = 0.$$

Three independent integrals which do not contain time explicitly are given by the functions

$$\varphi', \psi', \varphi\psi' - \psi\varphi'.$$

The first two of these integrals are normal (since two physically equivalent points can differ by integer values of the variables φ and ψ but φ' and ψ' will have for them the same values). The third integral is not normal. Indeed, let I be the value of this integral for some state of the system (φ, ψ, φ', ψ'). For any integers k and l the point ($\varphi + k$, $\psi + l$, φ', ψ') represents the same state of the system, but the value of the third integral at this point is $I + k\psi' - l\varphi'$, which in general is different from I. Furthermore, if the values φ' and ψ' are incommensurable, the third integral assumes an everywhere dense set of values for the same state of the system.

According to what was said above, if we desire to construct a reduced manifold metrically indecomposable in the extended sense, we have to fix the values of the integrals φ' and ψ'. Let $\varphi' = \alpha$, $\psi' = \beta$, where α and β are any two incommensurable real numbers. The reduced space (φ, ψ) will be then a two-dimensional part of the four-dimensional phase space, and will be also a part of the surface of constant energy $E = (1/2)(\alpha^2 + \beta^2)$. Let us investigate whether this plane (φ, ψ) admits of normal decompositions.

Since every square

$$k \leq \varphi < k + 1, \qquad l \leq \psi < l + 1, \tag{22}$$

where k and l are any integers, is physically equivalent to any other such square, and since in a normal subdivision all physically equivalent points must belong to the same part for each normal subdivision of the plane (φ, ψ) all such squares will be subdivided in parts which are mutually congruent.

A little explanation is necessary here: the sets M_1 and M_2 which normally subdivide the plane (φ, ψ) clearly will have infinite measure, which is not foreseen by the definition of a normal subdivision. In our case, however, this cannot cause any difficulty, since, in view of the physical equivalency of any two squares of the type (22), to take the phase average of any normal phase function we could restrict ourselves to the consideration of the fundamental square $0 \leq \varphi < 1, 0 \leq \psi < 1$. If we transfer any normal subdivision of this square to all squares (22), we obtain a subdivision of the plane (φ, ψ), which naturally may be called a normal subdivision of the plane (φ, ψ).

Let (M_1, M_2) be any such normal subdivision. Consider the side $\varphi = 0$ of the fundamental square. Let the point $(0, b)$ $(0 < b < 1)$ of this side belong to the set M_1. We assert that in this case the point $(0, \rho(b + k\beta/\alpha))$,[4] where k is any integer, also belongs to M_1. Indeed the set M_1, being invariant, contains together with the front $(0, b)$ the whole trajectory passing through this point, that is, all points of the type $(\alpha t, b + \beta t)$, and, in particular (for $t = k/\alpha$) the point $(k, b + k\beta/\alpha)$. But this point is physically equivalent to the point $(0, \rho(b + k\beta/\alpha))$ of the fundamental square, and therefore, since our subdivision is normal, must also belong to M_1. The numbers $\rho(b + k\beta/\alpha)(-\infty < k < \infty)$, if α and β are incommensurable, constitute an everywhere dense set, hence the set M_1, is everywhere dense on the side $\varphi = 0$ of our square. Let A_1 be the set which is common to M_1 and this side. We can assert that if A_1 is

[4]Here $\rho(x) = x - [x]$ is the "fractional part" of the real number x.

measurable, it must have positive (linear) measure. Indeed, if the measure of A_1 were equal to zero, then the part of M_1 in our square which consists of the family of parallel lines of slope β/α, passing through the points of A_1 would have been of plane measure zero. In view of the mutual congruency of subdivisions induced in all squares, we would have $\mathfrak{M}M_1 = 0$, contrary to our assumption.

Let $\epsilon > 0$ be arbitrarily small, and $\delta(b_1, b_2)$ such an interval on the side $\varphi = 0$ of the fundamental square, in which the average density of the set A_1 is greater than $(1 - \epsilon)$, that is

$$\mathfrak{M}(\delta \cdot A_1) > (1 - \epsilon)\mathfrak{M}\delta$$

(such an interval exists in view of a known theorem concerning the density of measurable sets). After the time $t = k/\alpha$ the interval δ passes over into an interval of the same length on the line $\varphi = k$, which, in its turn, is equivalent to some interval (or a pair of intervals) δ' of the side $\varphi = 0$ of our square, and we must have

$$\mathfrak{M}\delta' = \mathfrak{M}\delta.$$

Since this set M_1 is invariant,

$$\mathfrak{M}(\delta \cdot A_1) = \mathfrak{M}(\delta' \cdot A_1)$$

so that

$$\mathfrak{M}(\delta' \cdot A_1) > (1 - \epsilon)\mathfrak{M}\delta'.$$

Varying k we obtain, as it is easy to see, along the side $\varphi = 0$ of our square a dense (because of the irrationality of β/α) set of intervals δ' of equal length within which the mean density of the set A_1 exceeds $1 - \epsilon$. It follows that:

$$\mathfrak{M}A_1 \geq 1 - \epsilon$$

or, because of the arbitrary value of ϵ:

$$\mathfrak{M}A_1 = 1.$$

Writing A_2 for a set complimentary to A_1 (on the side $\varphi = 0$) we have:

$$\mathfrak{M}A_2 = 0.$$

As we have already seen this would lead to $\mathfrak{M}M_2 = 0$ which contradicts our assumption. This argument shows that the plane (φ_1, φ_2) is metrically indecomposable in the general sense of this word. The question as to whether this metric indecomposability can be considered as the general property of the broad class of systems encountered in statistical physics cannot be answered at the present time. We notice however that many other authors succeeded in the construction of rather general examples of the above given type and gave arguments in favor of the generality of the above statement. We will not enter here into the discussion of these problems, but will turn to the analysis of those cases which are of greatest importance in statistical mechanics.

13. The possibility of a formulation without the use of metric indecomposability. All the results obtained by Birkhoff and his followers (as well as all considerations of the previous section) pertain to the most general type of dynamic systems, and consider different problems connected with them. The authors of these studies have been working, as a rule, on the development of the so-called "general dynamics"—an important and interesting branch of modern mechanics. They have not been interested in the problem of the foundation of statistical mechanics which is our primary interest in the present book. Their aim was to obtain the results in the most general from; in particular all these results pertain equally to the systems with only few degrees of freedom as well as to the systems with a very large number of degrees of freedom.

From our point of view we must deviate from this tendency. We would unnecessarily restrict ourselves by neglecting the special properties of the systems considered in statistical mechanics (first of all their fundamental property of having a very large number of the degrees of freedom), and demanding the applicability of the obtained results to any dynamic system. Furthermore, we do not have any basis for demanding the possibility of substituting phase averages for the time averages of *all* functions; in fact the functions for which such substitution is desirable have many specific properties which make such a substitution apparent in these cases. In the present section we shall make several elementary remarks along these lines.

In the field of statistical mechanics we are, first of all, helped by the fact that the majority of phase functions describing the most important physical quantities exhibit a very peculiar behavior (compare section 10). In fact these functions are, as a rule, approximately constant on the surfaces of constant energy, i.e., with the exception of a set of points of a very small measure, they possess on each such surface values which are very close to a certain number characteristic of the surface. The reasons for such peculiar behavior will be partially discussed later in this chapter, and we will return to them in more detail in the later chapters. We will remark here, however, that these reasons arise partially from the peculiar properties of mechanical systems treated in statistical physics (breaking up into a large number of components), and partially from the specific properties of the functions with which we are dealing (these are, as a rule, the "sum-functions", i.e., the sums of functions each depending on the dynamical coordinates of only one component). It is clear without calculation that, for such functions, the time averages taken along most trajectories must be very close to the corresponding phase averages. If derivable, however, the approximate proof of the above statement can be given along the following lines:

Let us assume that the values of the function $f(P)$ on the surface Σ_a (except in a set of points of a very small measure) are very close to a certain number A. Then, unless $f(P)$ assumes at this small set of points some particularly large values, the quantity

$$\frac{1}{\Omega(a)} \int_{\Sigma_a} | f(P) - A | \frac{d\Sigma}{\operatorname{grad} E} = I$$

will in general be small. Assume for simplicity $A = 0$, which apparently does not reduce the generality of our considerations. Assume also, as we did before, that

$$\frac{1}{C}\int_0^C f(P, t)\,dt = f_C(P), \qquad \lim_{C\to\infty} f_C(P) = \hat{f}(P);$$

finally let M_α be a set of points on the surface Σ_a for which $|\hat{f}(P)| > \alpha$, and M_α^C a set of points for which $|f_C(P)| > \alpha/2$. Since under the condition $C\to\infty$ (on the surface Σ_a), $f_C(P)\to \hat{f}(P)$, we can write for sufficiently large C:

$$\mathfrak{M}M_\alpha^C > \tfrac{1}{2}\mathfrak{M}M_\alpha \text{ [5]}$$

consequently:

$$\begin{aligned}\frac{\alpha\mathfrak{M}M_\alpha}{4} &< \int_{M^C_\alpha} |f_C(P)|\frac{d\Sigma}{\operatorname{grad} E}\\ &\le \frac{1}{C}\int_0^C dt\int_{M^C_\alpha} |f(P, t)|\frac{d\Sigma}{\operatorname{grad} E}\\ &= \frac{1}{C}\int_0^C dt\int_{M^C_\alpha(t)} |f(P)|\frac{d\Sigma}{\operatorname{grad} E}\\ &\le \frac{1}{C}\int_0^C dt\int_{\Sigma_a} |f(P)|\frac{d\Sigma}{\operatorname{grad} E} = I\Omega(a),\end{aligned}$$

from which follows:

$$\frac{\mathfrak{M}M_\alpha}{\Omega(a)} < \frac{4I}{\alpha};$$

if for example we choose $\alpha = I^{1/2}$ we obtain:

$$\frac{\mathfrak{M}M_{I^{1/2}}}{\Omega(a)} < 4(I)^{1/2}$$

---

[5] To prove this inequality assume that $\overline{M_\alpha^C}$ is a set complementary to M_α^C. If $P\in M_\alpha$ and $P\in\overline{M_\alpha^C}$, we evidently have: $|f_C(P) - \hat{f}(P)> \alpha/2$. Hence, because of convergence in measure under the condition $C\to\infty$, we have $\mathfrak{M}(M_\alpha\cdot\overline{M_\alpha^C})\to 0$, or $\mathfrak{M}(M_\alpha\cdot M_\alpha^C)\to\mathfrak{M}M_\alpha$. From this it follows that, for sufficiently large C, $\mathfrak{M}(M_\alpha\cdot M_\alpha^C) > \frac{1}{2}\mathfrak{M}M_\alpha$, or *a fortiori* $\mathfrak{M}M_\alpha^C > \frac{1}{2}\mathfrak{M}M_\alpha$.

so that the relative measure of the set of points on the surface Σ_a for which $| \hat{f}(P) |$ exceeds the small quantity $I^{1/2}$ is smaller than the small quantity $4(I)^{1/2}$. It is clear that in order to reach the practical conclusions from the above calculation, we must estimate in each particular case the order of magnitude of the small quantity I. In many cases such an estimate is actually possible. However, it is also possible to make some estimates of quite a general nature. Thus, for example, we will see in the following chapters that for a physical system formed by n molecules, the most important phase functions are of the order of magnitude n. The "dispersion" of such a function, i.e., the quantity

$$I' = \frac{1}{\Omega(a)} \int_{\Sigma_a} [f(P) - A]^2 \frac{d\Sigma}{\text{grad } E},$$

has also, as a rule, the order of magnitude n (Chapter VIII, Sec. 36). Since, because of the Schwartz inequality $I \leq (I')^{1/2} = 0(n^{1/2})$, we find, choosing $\alpha = I^{3/2}$ (order of magnitude $n^{3/4}$), that the relative measure of the set of points for which

$$| \hat{f}(P) - A | > Kn^{\frac{3}{4}}$$

or

$$\left| \frac{\hat{f}(P)}{A} - 1 \right| > K_1 n^{-\frac{1}{4}}$$

is a small quantity of an order of magnitude not less than $I/\alpha = 0(n^{-1/4})$ (K and K_1 being positive constants). Since the quantity A can be assumed to represent the phase average $\bar{f}$ of the function $f(P)$ on the surface Σ_a, the above considerations supply certain approximate qualitative estimates pertinent to the substitution of phase averages by time averages. These almost trivial considerations lead us to suppose that, at least in the fundamental problems of statistical mechanics, and especially for practical purposes, we can avoid the use of the ergodic theorems of general dynamics.

We make one more remark. In this as well as in the previous sections, we have been satisfied with such situations in which the desirable phenomenon was taking place at all points of the surface Σ_a except at a set of points with a very small measure (sometimes exactly zero). It is clear that, taking this point of view, we make the definite assumption that, if some collection of the states of the system is represented on the surface Σ_a by a set of points of a very small relative measure, then the states belonging to this collection appear very infrequently in practice.

The exact mathematical formulation of this assumption in terms of the theory of probability is as follows: considering the different states of the system (i.e., of the points of the surface Σ_a) as random events, we assume that they are subject to any, not necessarily *absolutely continuous*, distribution law (i.e., such a distribution law for which the collection of very small measure possesses a very small probability). Such an assumption is in fact absolutely unavoidable in any comparison of our theory with reality. As a working hypothesis it is quite natural and gives us a free hand in selecting free distributions appropriate in practice.

Let us consider, finally, one more simple argument pertinent to the same group of ideas. Let us call a summable function $f(P)$ ergodic if, for almost all trajectories, $\hat{f}(P) = \bar{f}$. As remarked before, most of the phase functions considered in statistical mechanics are of the "summable type" i.e. represent the sum of functions each one depending only on the coordinates of a single molecule. It is clear that such a function will be ergodic if each of its components is also ergodic, since the averages $\hat{f}$ and $\bar{f}$ are both linear transformations. Thus, to prove the ergodic nature of such functions, it suffices to prove it for the functions corresponding to single molecules. We will give now some considerations in favor of the above statement.

Let $f(P)$ be a function of coordinates of only one molecule. Without restricting the generality of the argument, assume $\bar{f} = 0$. Let M be the upper limit of the function $|f|$ on the surface Σ_a, and

$$Df = \overline{f^2(P)},$$

$$R(u) = \frac{1}{Df} \overline{f(P, t)f(P, t + u)}$$

(it is apparent that the last quantity is independent of t). Thus, Df is the phase dispersion of the function $f(P)$, and $R(u)$ is the phase coefficient of correlation connecting $f(P, t)$ and $f(P, t + u)$. Because of the fact that the given system consists of a very large number of molecules it is natural to expect that knowledge of the state of a single molecule at a certain moment does not permit us to predict anything (or almost anything) about the state in which this molecule will be found after a sufficiently long time. For example, the exact knowledge of the energy of a given molecule at a given moment of time cannot give us any indications concerning the value which this energy will have several hours later (because of the large numbers of collisions suffered by the molecule during this time interval). This statement seems to us so natural that it would be difficult to think otherwise; in fact, this represents the basic idea of "molecular chaos". Expressing this in terms of the theory of probabilities we can say that stochastic dependence between the quantities $f(P, t)$ and $f(P, t + u)$ decreases very rapidly with increasing u, and almost entirely vanishes for appreciably large values of u; in particular it means that $R(u)$ must be small for large u and that $R(u) \to 0$ for $u \to \infty$. The only bad point of the above argument which must be mentioned here is that, since $R(u)$ is the *phase* coefficient of the correlation, we cannot be sure to what extent it can be used to characterize the stochastic dependence between the quantities in question. However, the relation $R(u) \to 0$ $(u \to \infty)$ represents a well defined property of the function $f(P)$ and of the natural motion on the surface Σ_a, —a property which must necessarily take place if the initial correlation between the quantities $f(P, t)$ and $f(P, t + u)$ becomes, "generally speaking", nearer with increasing u. The expression "generally speaking" attains exact meaning in terms of the measure on the surface Σ_a, and, as we have seen above, the stochastic interpretation of this measure represents the necessary postulate of the entire theory.

From this point of view it is interesting to prove the following theorem:

Theorem: If $R(u) \to 0$ for $u \to \infty$ the function $f(P)$ is ergodic.

Proof: Assuming (as usual)

$$\lim_{C\to\infty} \int_0^C f(P, t)\, dt = \hat{f}(P)$$

we have

$$\overline{\hat{f}^2(P)} = \frac{1}{\Omega(a)} \int_{\Sigma_a} \frac{d\Sigma}{\text{grad } E} \hat{f}^2(P)$$

$$= \frac{1}{\Omega(a)} \int_{\Sigma_a} \frac{d\Sigma}{\text{grad } E} \left\{ \hat{f}^2(P) - \frac{1}{C^2} \int_0^C \int_0^C f(P, u) f(P, v)\, du\, dv \right\}$$

$$+ \frac{1}{\Omega(a)} \int_{\Sigma_a} \frac{d\Sigma}{\text{grad } E} \cdot \frac{1}{C^2} \int_0^C \int_0^C f(P, u) f(P, v)\, du\, dv.$$

Writing Q for the above expression in brackets, and G_ϵ and $\overline{G}_\epsilon$ respectively for the set of points on the surface Σ_a for which $|Q| < \epsilon$ and its complementary set

$$\overline{\hat{f}^2(P)} = \frac{1}{\Omega(a)} \frac{1}{C^2} \int_0^C \int_0^C du\, dv \int_{\Sigma_a} \frac{f(P, u) f(P, v)\, d\Sigma}{\text{grad } E}$$

$$+ \frac{1}{\Omega(a)} \int_{G_\epsilon} \frac{Q\, d\Sigma}{\text{grad } E} + \frac{1}{\Omega(a)} \int_{\overline{G}_\epsilon} \frac{Q\, d\Sigma}{\text{grad } E}$$

since

$$Q = \hat{f}^2(P) - \left\{ \frac{1}{C} \int_0^C f(P, t)\, dt \right\}^2$$

we conclude, that under the condition $C \to \infty$, $Q \to 0$ almost everywhere on the surface Σ_a. Consequently $\mathfrak{M}\overline{G}_\epsilon \to 0$ from which follows that for sufficiently large C:

$$\frac{\mathfrak{M}\overline{G}_\epsilon}{\Omega(a)} = \frac{1}{\Omega(a)} \int_{\overline{G}_\epsilon} \frac{d\Sigma}{\text{grad } E} < \epsilon$$

since obviously

$$\frac{1}{\Omega(a)} \int_{\Sigma_a} \frac{f(P, u) f(P, v)\, d\Sigma}{\text{grad } E} = Df\, R(u - v)$$

and

$$| Q | < \epsilon \qquad (P \in G_\epsilon)$$

$$| Q | \leq M^2 \qquad (P \in \overline{G}_\epsilon)$$

we conclude that:

$$\overline{\hat{f}^2(P)} \leq \left| \frac{Df}{C^2} \int_0^C \int_0^C R(u - v)\, du\, dv \right| + \epsilon + M^2 \epsilon.$$

If $| R(a) | < \epsilon$ for $| a | > a_0$ we get (taking into consideration that $| R(a) | \leq 1$ for any a)

$$\overline{\hat{f}^2(P)} \leq \frac{Df}{C^2} \int_0^C du \int_{max(0, u - a_0)}^{min(C, u + a_0)} | R(u - v) |\, dv$$

$$+ \frac{\epsilon Df}{C^2} \int_0^C \int_0^C du\, dv + \epsilon + M^2 \epsilon$$

$$\leq \frac{2a_0 Df}{C} + \epsilon(Df + 1 + M^2).$$

Since we can choose ϵ arbitrarily small and C arbitrarily large:

$$\overline{\hat{f}^2(P)} = 0$$

so that almost everywhere on Σ_a

$$\hat{f}(P) = 0 = \overline{f}.$$

This proves the ergodic nature of the function of $f(P)$.

CHAPTER IV

REDUCTION TO THE PROBLEM OF THE THEORY OF PROBABILITY

14. Fundamental distribution law. We shall now consider the aggregate of $2s$ dynamic variables $(x_1, x_2, \cdots, x_{2s})$, determining the state of a given system G with s degrees of freedom, as a multidimensional probability quantity (probability vector). We shall assume as usual that the energy E of the system has a certain constant value a, so that all possible values of the probability vector $(x_1, x_2, \cdots, x_{2s})$ correspond to the points on a certain surface Σ_a. The probability that the representative point of the system will fall within a certain set M on the surface Σ_a will be assumed to be given by:

$$\frac{1}{\Omega(a)} \int_M \frac{d\Sigma}{\operatorname{grad} E},$$

where the value $\Omega(a) = \int_{\Sigma_a} d\Sigma/\operatorname{grad} E$ of the surface Σ_a is the structure function of the system. It is obvious that the probability field introduced in this way satisfies all necessary conditions. The distribution law of the probability vector $(x_1, x_2, \cdots, x_{2s})$ thus established will be called in the future *the fundamental distribution law of the system (for $E = a$).* Let $\varphi(x_1, x_2, \cdots, x_{2s})$ be an arbitrary measurable phase function of the system G. Then the probability of the inequality

$$\varphi(x_1, x_2, \cdots, x_{2s}) < x,$$

where x is an arbitrary real number, will be determined by the formula

$$\mathbf{P}(\varphi < x) = \frac{1}{\Omega(a)} \int_{\varphi < x} \frac{d\Sigma}{\operatorname{grad} E}.$$

Thus, any measurable phase function can be considered as an

accidental quantity with a well defined distribution law. The mathematical expectation of this quantity,

$$\mathbf{E}\varphi = \frac{1}{\Omega(a)} \int_{\Sigma_a} \varphi(x_1, \cdots, x_{2s}) \frac{d\Sigma}{\operatorname{grad} E},$$

coincides (under the assumption of absolute convergence of this integral) with the quantity which we called the phase average $\bar{\varphi}$ of the function φ in section 7 of Chapter II. In particular, if φ is an ergodic function, its time average for almost any trajectory on the surface Σ_a coincides with its mathematical expectation $\mathbf{E}\varphi = \bar{\varphi}$.

If φ is a function characteristic of a certain measurable set of points M on the surface Σ_a (i.e. if $\varphi = 1$ within M and $\varphi = 0$ without), $\mathbf{E}\varphi$ obviously gives the probability of finding the point $(x_1, x_2, \cdots, x_{2s})$ within the set of points M. If φ is an ergodic function this probability coincides with the relative mean time spent by the moving point within the set M for almost any trajectory located on the surface Σ_a.

The fundamental distribution law formulated above permits us to introduce convenient probabilistic terminology for the ideas connected with evaluation of phase averages. At the same time, as we shall see later, this formulation of the fundamental distribution law will permit us to use the well known analytical apparatus of the theory of probability for the solution of many fundamental problems in statistical mechanics.

15. The distribution law of a component and its energy. Let a given system G have a component G_1 with dynamic coordinates $(x_1, x_2, \cdots, x_r)$ (the complementary component G_2 having dynamic coordinates $x_{r+1}, \cdots, x_{2s}$). The fundamental distribution law assumed for the system G, i.e. for the multidimensional random quantity $(x_1, \cdots, x_{2s})$, uniquely determines, according to the well known rules of probability, the distribution law for the arbitrary group of dynamical variables in the system G_1.

In particular, the set of variables $(x_1, x_2, \cdots, x_r)$ $(r < 2s)$ or, as we shall say for brevity, the component G_1 is subject to a definite distribution law in the space of r dimensions, which, of course, coincides with its phase space. Let us now find this distribution law.

Let M_1 be a measurable set in the phase space Γ_1 of the component G_1, in which $(x_1, x_2, \cdots, x_r)$ serve as Cartesian coordinates. Further, let M be a set of points in the phase space Γ of the system G for which the first r coordinates represent a point in the space Γ_1 belonging to the set M_1 (so that the point $(x_1, \cdots, x_r)$ will belong to the set M_1, when and only when the point $(x_1, \cdots, x_{2s})$ belongs to the set M). The probability that the representative point P_1 of the component G_1 falls within the set M_1 coincides with the probability that the representative point P of the component G falls within the set M (both probabilities being determined, as usual, under the assumption that $E = a$), so that we have (denoting by AB the intersection i.e the common part of the sets A and B):

$$\mathbf{P}(P_1 \in M_1) = \mathbf{P}(P \in M) = \frac{1}{\Omega(a)} \int_{M\Sigma_a} \frac{d\Sigma}{\text{grad } E}$$

$$= \frac{1}{\Omega(a)} \int_{\Sigma_a} \varphi \frac{d\Sigma}{\text{grad } E},$$

where φ stands for the previously defined function characteristic of the set M. Because of the general theorem (section 7, Chapter II) this gives us:

$$\mathbf{P}(P_1 \in M_1) = \frac{1}{\Omega(a)} \frac{d}{da} \int_{V_a} \varphi \, dV, \tag{23}$$

where dV stands for the volume element in the phase space Γ of the system G i.e.

$$dV = dx_1 \cdots dx_{2s} .$$

Since the function φ is independent of the variables $x_{r+1}, \cdots, x_{2s}$ we conclude that

$$\int_{V_a} \varphi \, dV = \int_{\Gamma_1} \varphi \, dV_1 \int (V_{a-E_1})_2 \, dV_2 ,$$

where $dV_1 = dx_1 \cdots dx_r$, $dV_2 = dx_{r+1} \cdots dx_{2s}$, $E_1 = E_1(x_1, \cdots, x_r)$ and $E_2 = E_2(x_{r+1}, \cdots, x_{2s})$ represent the energies of the components G_1 and G_2, whereas $(V_{a-E_1})_2$ is the set of points in the space Γ_2 for which $E_2 < a - E_1$. The outer integration can be extended through the entire space Γ_1 without changing the results since for $E_1 > a$ the inner integral vanishes.

In the above expression the inner integral represents the volume of that part of the phase space Γ_2 of the component G_2 where $E_2 < a - E_1$. Denoting it as usual by $V_2(a - E_1)$ we will have:

$$\int_{V_a} \varphi\, dV = \int_{\Gamma_1} \varphi V_2(a - E_1)\, dV_1 = \int_{M_1} V_2(a - E_1)\, dV_1,$$

consequently:

$$\frac{d}{da} \int_{V_a} \varphi\, dV = \int_{M_1} \Omega_2(a - E_1)\, dV_1,$$

since it is clear from the definition of the structure function that $V_2'(x)$ coincides with the structure function $\Omega_2(x)$ of the component G_2. Thus the relation (23) gives us

$$\text{(24)} \qquad \mathbf{P}(P_1 \in M_1) = \frac{1}{\Omega(a)} \int_{M_1} \Omega_2(a - E_1)\, dV_1.$$

It must be remembered that in the above expression $dV_1 = dx_1 \cdots dx_r$ and E_1 is the function of $x_1, \cdots, x_r$.

We see that in the case where the energy of the system G is equal to a, the distribution law of the component G_1 in its phase space Γ_1 is given by the density function

$$\text{(25)} \qquad \frac{\Omega_2(a - E_1)}{\Omega(a)},$$

where $\Omega_2(x)$ is the structure function of the complementary component G_2. This fact permits us to write the expression for the phase average of any function depending on $x_1, \cdots, x_r$, in the form of an integral extended over the space Γ_1.

Indeed, let $\varphi(x_1, \cdots, x_r)$ be such a function. We know that its phase average $\bar{\varphi}$ coincides with the mathematical expectation $\mathbf{E}\varphi$ which, according to the above results, can be written in the form

$$(26) \qquad \bar{\varphi} = \mathbf{E}\varphi = \frac{1}{\Omega(a)} \int_{\Gamma_1} \varphi \Omega_2(a - E_1)\, dV_1 \, .$$

The most important function of the above type is the energy $E_1 = E_1(x_1, \cdots, x_r)$ of the component G_1. Because of (26) we have:

$$\bar{E}_1 = \mathbf{E}E_1 = \frac{1}{\Omega(a)} \int_{\Gamma_1} E_1 \Omega_2(a - E_1)\, dV_1 \, .$$

However, because of the particular importance of the quantity E_1, we will not limit ourselves by establishing only its mean value, but we will also find its distribution law.

We have seen that the aggregate $(x_1, \cdots, x_r)$ of the dynamical coordinates of the component G_1 is a multi-dimensional random quantity distributed in the space Γ_1 with the density

$$\frac{\Omega_2(a - E_1)}{\Omega(a)}.$$

Accordingly, the probability that $g_1 < E_1 < g_2$ is given by:

$$\mathbf{P}(g_1 < E_1 < g_2) = \frac{1}{\Omega(a)} \int_{g_1 < E_1 < g_2} \Omega_2(a - E_1)\, dV_1 \, .$$

According to formula (18) Chapter II, this multiple integral in the space of r dimensions can be written in the form of a simple integral

$$\int_{g_1}^{g_2} \Omega_1(E_1)\, \Omega_2(a - E_1)\, dE_1 \, ,$$

which brings us to the relation

$$\mathbf{P}(g_1 < E_1 < g_2) = \frac{1}{\Omega(a)} \int_{g_1}^{g_2} \Omega_1(x)\, \Omega_2(a - x)\, dx.$$

Thus the random quantity E_1 is subject to probability density

$$\frac{\Omega_1(x)\,\Omega_2(a - x)}{\Omega(a)}. \tag{27}$$

This permits us to express the phase average of any function $\varphi(E_1)$ of the energy of the component G_1 in the form of an ordinary integral:

$$\bar{\varphi} = \mathbf{E}\varphi(E_1) = \int \varphi(x)\,\frac{\Omega_1(x)\,\Omega_2(a - x)}{\Omega(a)}\,dx. \tag{28}$$

In particular:

$$\bar{E}_1 = \mathbf{E}E_1 = \frac{1}{\Omega(a)} \int x\Omega_1(x)\,\Omega_2(a - x)\,dx. \tag{29}$$

In the last two formulae the integrals can be taken between infinite limits; in fact, since the integrated function is different from zero only for $0 < x < a$ there is no divergence difficulty.

In applications we will usually encounter phase functions which depend on the dynamical coordinates of some components of the given system, and include essentially the energy of this component. As we have just seen, the distribution law for the energy of the given component as well as for its dynamic variables contains the structure functions Ω, Ω_1, and Ω_2. (It may be noted that the general formulae determining the mean values of an arbitrary phase function on the surface Σ_a also contain the quantity $\Omega(a)$). Thus it is clear that any analytical method of deriving the approximate formulae for the mean values of the phase function used in statistical mechanics must first of all give convenient approximate expressions for the structure functions. Accordingly, in our approach to the problem, we will try to use the fact that the systems usually considered in statistical mechanics consist of a very large number of similar components. Using the methods of the theory of probability we will be able to establish for the structure functions of such systems the approximate expressions which are to a large extent independent of the nature of individual components.

16. Generating functions. Let us consider a system G whose structure function $\Omega(x)$ is subject to the usual conditions: it is positive and monotonically increasing for $x > 0$, it is equal to zero for $x \leq 0$, it is continuous and increases without bound for $x \to \infty$. However, we will require the integral

$$\Phi(\alpha) = \int e^{-\alpha x}\Omega(x)\, dx \tag{30}$$

to converge for any $\alpha > 0$ (it may be remarked that this condition is satisfied in all actual physical problems).

In the future we will call the function $\Phi(\alpha)$, (which is nothing else but the "Laplace transform" of the structure function $\Omega(x)$), the generating function of the system G, because of its fundamental role in our analytical method. For the same reason we will discuss in more detail the fundamental properties of such generating functions.

Each generating function is completely determined for all positive values of its argument by the expression (30); only this case will be considered. From the definition of the generating function it follows that:

(1) $\Phi(\alpha)$ *is a positive and monotonically decreasing function of* α.
(2) $\Phi(\alpha) \to \infty$ *for* $\alpha \to 0$.

Furthermore, it is easy to prove that:

(3) *For any* $\alpha > 0$, $\Phi(\alpha)$ *has derivatives of all orders. For* $n = 1, 2, \cdots$:

$$\Phi^{(n)}(\alpha) = (-1)^n \int x^n e^{-\alpha x}\Omega(x)\, dx. \tag{31}$$

In fact, for any positive number α_0, and for any large number n, we have, for sufficiently large x and $\alpha \geq \alpha_0$:

$$x^n e^{-\alpha x} < e^{\alpha_0 x/2} e^{-\alpha_0 x} = e^{-\alpha_0 x/2}.$$

From this follows that the integral in the expression (31) converges uniformly for $\alpha \geq \alpha_0$.

We also notice that, since $\Phi(\alpha)$ is always positive, the function $\log \Phi(\alpha)$ also possesses all the properties mentioned above (except, of course, being positive); in particular any generating function has logarithmic derivatives of all orders.

(4) *The second logarithmic derivative of the function* $\Phi(\alpha)$ *is always positive for* $\alpha > 0$.

In fact direct calculation shows that:

$$\frac{d^2 \log \Phi(\alpha)}{d\alpha^2} = \frac{\Phi(\alpha)\Phi''(\alpha) - [\Phi'(\alpha)]^2}{[\Phi(\alpha)]^2}$$

$$= \frac{1}{\Phi(\alpha)} \int \left(x + \frac{\Phi'(\alpha)}{\Phi(\alpha)}\right)^2 e^{-\alpha x}\Omega(x)\, dx > 0.$$

From this follows:

(5) *The equation*

$$-\frac{\Phi'(\alpha)}{\Phi(\alpha)} = a$$

has one single positive solution for any $a > 0$.

In fact consider the function

$$\Phi_a(\alpha) = e^{a\alpha}\Phi(\alpha).$$

Because of the property (2) $\Phi_a(\alpha) \to \infty$ for $\alpha \to 0$, and since

$$\Phi_a(\alpha) > e^{a\alpha} \int_0^{a/2} e^{-\alpha x}\Omega(x)\, dx > e^{a\alpha/2} \int_0^{a/2} \Omega(x)\, dx,$$

we can conclude that $\Phi_a(\alpha) \to \infty$ also for $\alpha \to \infty$. It is also apparent that the function $\log \Phi_a(\alpha)$ possesses the same properties. However, $\log \Phi_a(\alpha)$ is convex function since its second derivative, coinciding with the derivative of $\log \Phi(\alpha)$, is always positive because of the property (4); this shows that the function $\log \Phi_a(\alpha)$, becoming infinite for $\alpha \to 0$ and $\alpha \to \infty$, must necessarily possess a single minimum. In the point of minimum

$$\frac{d \log \Phi_a(\alpha)}{d\alpha} = a + \frac{\Phi'(\alpha)}{\Phi(\alpha)} = 0$$

which proves our statement.

The most important property of generating functions is their law of composition, i.e. the law by which the generating function is constructed from the generating functions of its components. Let the system G consist of two components G_1 and G_2 with the structure functions $\Omega_1(x)$ and $\Omega_2(x)$ and the generating function $\Phi_1(\alpha)$ and $\Phi_2(\alpha)$. Since, according to the formula (20) section 8, Chapter II,

$$\Omega(x) = \int \Omega_1(x - y)\,\Omega_2(y)\,dy,$$

we have

$$\Phi(\alpha) = \int e^{-\alpha x}\Omega(x)\,dx = \int e^{-\alpha x}\,dx \int \Omega_1(x - y)\,\Omega_2(y)\,dy$$

$$= \int \Omega_2(y)e^{-\alpha y}\,dy \int \Omega_1(x - y)e^{-\alpha(x-y)}\,dx$$

$$= \int \Omega_2(y)e^{-\alpha y}\,dy \int \Omega_1(z)e^{-\alpha z}\,dz = \Phi_1(\alpha)\Phi_2(\alpha).$$

It is clear that, using the method of mathematical induction, we can generalize this result for the functions consisting of many components. Thus we come to the following rule:

(6) *The generating function of a system G is equal to the product of the generating functions of its components.*

Thus, for example, if G is a gas, consisting of n identical molecules and if $\varphi(\alpha)$ is the generating function of a single molecule we have

$$\Phi(\alpha) = [\varphi(\alpha)]^n.$$

If G is a mixture of two gases consisting of n_1 molecules with the generating function $\varphi_1(\alpha)$ and n_2 molecules with the generating function $\varphi_2(\alpha)$, we have:

$$\Phi(\alpha) = [\varphi_1(\alpha)]^{n_1}[\varphi_2(\alpha)]^{n_2}$$

etc.

Thus we see that for the composite mechanical system, generating functions are subject to a much simpler composition law than the structure functions. This particular property of generating functions makes them particularly convenient for the study of systems consisting of a large number of components.

We may also remark that, on the basis of the general formula (19) Chapter II, the generating function $\Phi(\alpha)$ of the system G can be expressed in the form

$$\Phi(\alpha) = \int_{\Gamma} e^{-\alpha E}\, dV,$$

where E is a total energy of the system G considered as a function of coordinates in the phase space Γ.

17. Conjugate distribution laws. Consider again the system G discussed in a previous section and use the same notation as before. Assume

$$U^{(\alpha)}(x) = \begin{cases} \dfrac{1}{\Phi(\alpha)}\, e^{-\alpha x}\Omega(x) & (x \geq 0), \\ 0 & (x \leq 0). \end{cases} \tag{33}$$

Since,

$$U^{(\alpha)}(x) \geq 0, \qquad \int U^{(\alpha)}(x)\, dx = 1,$$

we conclude that $U^{(\alpha)}(x)$ represents a probability density for any $\alpha > 0$. For different α's we obtain an entire family of distribution laws. It is clear that this family is completely determined by the structure of the system G. We will call these the family of distribution laws *conjugate* with the system G.

Conversely, the structure function $\Omega(x)$ can be found from any of the conjugate distribution functions $U^{(\alpha)}(x)$ by means of the formula

$$\Omega(x) = \Phi(\alpha) e^{\alpha x} U^{(\alpha)}(x). \tag{34}$$

The mathematical expectation and the dispersion of a quantity distributed according to the conjugate law $U^{(\alpha)}(x)$ can be expressed in a simple way through the generating function $\Phi(\alpha)$ and its derivatives. In fact:

(35)
$$a = \int xU^{(\alpha)}(x)\,dx = \frac{1}{\Phi(\alpha)}\int xe^{-\alpha x}\Omega(x)\,dx = -\frac{\Phi'(\alpha)}{\Phi(\alpha)} = -\frac{d\log\Phi(\alpha)}{d\alpha}.$$

Remembering the property (5) of the generating functions we deduce an important theorem:

Theorem: For any positive number a, one can always find among the conjugate functions $U^{(\alpha)}(x)$ only one function which has mathematical expectation a.

Furthermore, the dispersion corresponding to $U^{(\alpha)}(x)$ is given by the expression:

(36)
$$\int (x-a)^2U^{(\alpha)}(x)\,dx = \left[\int x^2U^{(\alpha)}(x)\,dx\right] - a^2 = \frac{1}{\Phi(\alpha)}\left[\int x^2e^{-\alpha x}\Omega(x)\,dx\right] - a_2 = \frac{\Phi(\alpha)\Phi''(\alpha) - [\Phi'(\alpha)]^2}{[\Phi(\alpha)]^2} = \frac{d^2\log\Phi(\alpha)}{d\alpha^2}.$$

Finally, the composition law of the structure functions

$$\Omega(x) = \int \Omega_1(x-y)\Omega_2(y)\,dy$$

together with the expression (34), and the composition law of the leading functions (where $U_1^{(\alpha)}(x)$ and $U_2^{(\alpha)}(x)$ represent the conjugate distribution functions of the components G_1 and G_2) give us:

$$\Phi(\alpha)e^{\alpha x}U^{(\alpha)}(x) = \int \Phi_1(\alpha)e^{\alpha(x-y)}U_1^{(\alpha)}(x - y)\Phi_2(\alpha)e^{\alpha y}U_2^{(\alpha)}(y)\,dy$$

$$= \Phi_1(\alpha)\Phi_2(\alpha)e^{\alpha x}\int U_1^{(\alpha)}(x - y)U_2^{(\alpha)}(y)\,dy$$

$$= \Phi(\alpha)e^{\alpha x}\int U_1^{(\alpha)}(x - y)U_2^{(\alpha)}(y)\,dy.$$

from which follows:

$$U^{(\alpha)}(x) = \int U_1^{(\alpha)}(x - y)U_2^{(\alpha)}(y)\,dy. \tag{37}$$

Using the method of mathematical induction, we can generalize this important formula for the case where the system G consists of any number n of components with conjugate distributions $U_1^{(\alpha)}(x)$, $U_2^{(\alpha)}(x)$, $\cdots$, $U_n^{(\alpha)}(x)$. We have

$$U^{(\alpha)}(x) = \int \prod_{k=1}^{n-1}\left\{U_k^{(\alpha)}(x_k)\,dx_k\right\}U_n^{(\alpha)}\left(x - \sum_{k=1}^{n-1} x_k\right). \tag{38}$$

This is the composition law, well known in the theory of probability; it allows one to express the distribution of the sum of n independent random quantities in terms of the probability densities of individual components. We are led to the following rule concerning the composition of the conjugate distributions: *The conjugate distribution law of a given system can be derived from the corresponding distributions of its n components in the same way as the distribution of a sum of n mutually independent random quantities is found from the distributions of individual terms.*

It is clear that the value of the parameter α is quite arbitrary, but must be the same for all systems in question.

18. Systems consisting of a large number of components. Consider a system G consisting of the components G_1, G_2, $\cdots$,

G_n, where n is a very large number. According to the formula (34) of the previous section:

$$\Omega(x) = \Phi(\alpha)e^{\alpha x}U^{(\alpha)}(x).$$

We will try now to obtain a convenient approximate expression for the function $\Omega(x)$. The above expression for this function contains, apart from the elementary function $e^{\alpha x}$, the generating function $\Phi(\alpha)$ and the conjugate distribution function $U^{(\alpha)}(x)$. As can be easily seen and as we shall soon prove, the presence of the function $\Phi(\alpha)$ does not lead to any difficulties because of the extremely simple composition laws (section 16) governing the generating functions; in fact we have already seen that $\Phi(\alpha)$, being independent of x, plays the role of only a constant factor in the expression for the function $\Omega(x)$. Thus, the principal difficulty in our problem consists in finding a convenient approximate expression for the conjugate distribution function $U^{(\alpha)}(x)$.

We are helped here by the analytical methods of the theory of probability. In the previous section we have seen that $U^{(\alpha)}(x)$ represents the probability density for the sum of n random quantities (n being in the present case a very large number). For such cases the limit theorems of the theory of probability supply us with simple, convenient, and rather exact analytical approximations, the form of which does not depend on the special nature of the laws governing the separate components. These laws have only a small number of parameters entering into the approximate expressions. Thus, not having detailed information concerning the structure of the separate components of the system G and basing our conclusions exclusively on the very large numbers of these components we can arrive at important conclusions concerning the properties of this system. This result is typical of any application of the theory of probability and demonstrates its principal advantage in the study of mass phenomena.

Let us remark that the value of the parameter α still remains entirely arbitrary, so that we can use this extra degree of freedom for the simplification of the future calculations.

Thus, rather than creating the special analytical formalism for the purposes of statistical mechanics we plan to use in all future calculations the conventional formalism of the theory of probability. The next chapter will be devoted to the discussion of the fundamental steps to be taken along this line.

CHAPTER V

APPLICATION OF THE CENTRAL LIMIT THEOREM

19. Approximate expressions of structure functions. The most convenient formulation of the so called "central limit theorem" of the theory of probability, which gives the approximate expression for the distribution law governing the sum of a large number of mutually independent random quantities, can be given in the following form:

Consider a sequence of mutually independent random quantities with probability densities $u_1(x)$, $u_2(x)$, $\cdots$, *and characteristic functions* $g_1(t)$, $g_2(t)$, $\cdots$, *so that*

$$g_k(t) = \int e^{itx} u_k(x)\, dx \qquad (k = 1, 2, \cdots).$$

Let

$$\left.\begin{aligned} \int x u_k(x)\, dx &= a_k , \\ \int (x - a_k)^2 u_k(x)\, dx &= b_k , \\ \int | x - a_k |^3 u_k(x)\, dx &= c_k , \\ \int (x - a_k)^4 u_k(x)\, dx &= d_k , \\ \int | x - a_k |^5 u_k(x)\, dx &= e_k \end{aligned}\right\} \qquad (k = 1, 2, \cdots)$$

and assume that the given distribution laws are subject to the following conditions:

(1) *The functions* $u_k(x)$ *are differentiable and there exists a constant L such that*

$$\int | u_k'(x) | \, dx < L \qquad (k = 1, 2, \cdots).$$

(2) *There exist positive constants* α *and* β ($\alpha < \beta$) *such that:*

$$\alpha < b_k < \beta, c_k < \beta, d_k < \beta, e_k < \beta \qquad (k = 1, 2, \cdots).$$

(3) *There exist positive constants* λ *and* τ *such that in the region* $| t | \leq \tau$:

$$| g_k(t) | > \lambda \qquad (k = 1, 2, \cdots).$$

(4) *For each interval* (c_1, c_2) $(c_1 c_2 > 0)$ *there exists a number* $\rho = \rho(c_1, c_2) < 1$ *such that for any t in the interval* (c_1, c_2):

$$| g_k(t) | < \rho \qquad (k = 1, 2, \cdots).$$

Let us put $A_n = \sum_{k=1}^{n} a_k$, $B_n = \sum_{k=1}^{n} b_k$ *and write* $U_n(x)$ *as the probability density of the sum of the first n random quantities. Then, for* $n \to \infty$:

$$U_n(x) = \frac{1}{(2\pi B_n)^{1/2}} \exp\left[- \frac{(x - A_n)^2}{2B_n} \right]$$

$$\text{(39)} \qquad + \begin{cases} O\left(\dfrac{1 + | x - A_n |}{n^{3/2}}\right) & \text{for } | x - A_n | < 2 \log^2 n \\ O\left(\dfrac{1}{n}\right) & \text{for all } x. \end{cases}$$

The proof of this theorem is given in the appendix, together with a more exact formulation which, however, is not necessary for the purposes of the present chapter.

As indicated at the end of the previous chapter, we must use the central limit theorem for estimating the conjugate distribution function $U^{(\alpha)}(x)$ of the given system G under the assumption that the system consists of a very large number of components $g_1, g_2, \cdots, g_n$, with the structure functions $\omega_1(x), \omega_2(x), \cdots, \omega_n(x)$, the generating functions $\varphi_1(\alpha), \varphi_2(\alpha)$,

$\cdots$, $\varphi_n(\alpha)$ and conjugate functions $u_1^{(\alpha)}(x)$, $u_2^{(\alpha)}(x)$, $\cdots$, $u_n^{(\alpha)}(x)$. Since the latter functions play the role here of the functions $u_k(x)$ in the formulation of the limit theorem, we must initially make sure that the conjugate functions for the actual physical systems satisfy the conditions assumed in the proof of the limit theorem. This, however, does not present any difficulty.

The point is, that the conditions imposed on the functions $u_k(x)$ in the limit theorem are equivalent to assuming the uniformity of one or another property which they describe. However, in statistical physics the separate components g_k (molecules, atoms etc.) are always either of the same kind (homogeneous substance) or of a small number of different kinds (a mixture of several homogeneous substances). Thus, the structure functions and consequently also the conjugate functions for these components form a set within which all elements are either identical or break up into a small number of groups of identical elements. It is clear that under such conditions each characteristic of the functions $u_k^{(\alpha)}(x)$ appears uniformly in the entire set.

Let us consider now the separate conditions prescribed by the limit theorem. The structure function $\omega_k(x)$, as well as its derivative, is usually an analytic function which does not increase faster than a certain power of x when $x \to \infty$; since:

$$u_k^{(\alpha)}(x) = \frac{1}{\varphi_k(\alpha)} e^{-\alpha x} \omega_k(x),$$

the condition (1) is always satisfied.

The functions $u_k^{(\alpha)}(x)$ obviously always possess finite moments of all orders, whereas the uniformity restrictions on these moments follow directly from the above general remarks; thus the condition (2) is also always satisfied. The situation with the conditions (3) and (4) pertaining to the characteristic functions is even simpler. In fact, the condition (3) demands that $g_k(t)$ does not become zero for sufficiently small t; this is a property common to any characteristic function. The condition (4) demands only that $g_k(t) \neq 1$ for $t \neq 0$; this also is a property common to the characteristic functions of any continuous distribution law.

Thus we see that the use of the central limit theorem for the estimates of the conjugate distributions of a mechanical system gives definite answers. Introducing into the formula (34) (section 17 of the previous chapter) the approximate expression of $U^{(\alpha)}(x)$ as given by the expression (39) of the present section, and taking into account the formulae (35) and (36) (section 17, chapter IV), we must let

$$A_n = -\frac{d \log \Phi(\alpha)}{d\alpha}, \qquad B_n = \frac{d^2 \log \Phi(\alpha)}{d\alpha^2}$$

and thus obtain:

$$\Omega(x)$$

$$(40)\quad = \Phi(\alpha)e^{\alpha x}\left[\frac{1}{\left\{2\pi \dfrac{d^2 \log \Phi(\alpha)}{d\alpha^2}\right\}^{1/2}} \exp\left(-\frac{\left(x + \dfrac{d \log \Phi(\alpha)}{d\alpha}\right)^2}{2\dfrac{d^2 \log \Phi(\alpha)}{d\alpha^2}}\right) + \left\{\begin{array}{l} 0\left(\dfrac{1 + |x - A_n|}{n^{3/2}}\right) \text{ for } \left|x + \dfrac{d \log \Phi(\alpha)}{d\alpha}\right| < 2 \log^2 n \\ 0\left(\dfrac{1}{n}\right) \text{ for all } x. \end{array}\right\}\right]$$

We shall use this formula as a starting point in all future calculations.

Let us now consider the choice of the parameter α which so far has been arbitrary. In all cases when we speak about some system G with the constant energy a and about the different components of this system we will choose for α the simple root (comp. 5, section 16, chapter IV) of the equation:

$$-\frac{d \log \Phi(\alpha)}{d\alpha} = a.$$

This value of the parameter α we will denote in the future by ϑ. We will also assume:

$$\left(\frac{d^2 \log \Phi(\alpha)}{d\alpha^2}\right)_{\alpha=\vartheta} = B.$$

Using these notations for the structure function $\Omega(x)$ of the basic system G, we obtain from the formula (40) for $\alpha = \vartheta$ the following expression:

$$\Omega(x) = \Phi(\vartheta)e^{\vartheta x}\left[\frac{\exp\left[-\frac{(x-a)^2}{2B}\right]}{(2\pi B)^{1/2}} + \begin{cases} 0\left(\frac{1+|x-a|}{n^{3/2}}\right) & \text{for } |x-a| < 2\log^2 n \\ 0\left(\frac{1}{n}\right) & \text{for all } x. \end{cases}\right] \tag{41}$$

In particular, for $x = a$ this gives an important formula:

$$\Omega(a) = \Phi(\vartheta)e^{\vartheta a}\left[\frac{1}{(2\pi B)^{1/2}} + 0(n^{-3/2})\right]. \tag{42}$$

This formula gives the approximate expression corresponding to the surface of constant energy Σ_a in terms of the generating function $\Phi(\alpha)$ and its second logarithmic derivative for $\alpha = \vartheta$ where ϑ is determined by the basic relation

$$\left[-\frac{d \log \Phi(\alpha)}{d\alpha}\right]_{\alpha=\vartheta} = a.$$

20. A small component and its energy. Boltzmann's law. We have seen in section 15, Chapter IV that for the system G with the constant energy a which can be split into two components G_1 and G_2, the distribution law of the separate component G_1 in its phase space has the probability density

$$\frac{\Omega_2(a - E_1)}{\Omega(a)}, \tag{43}$$

where $\Omega_2(x)$ is the structure function of the component G_2 and E_1 is that function of the dynamic coordinates of the component G_1 which expresses its energy at the appropriate point of its phase space. Let us now assume, as is usually the case in the statistical mechanics, that the system G is composed of a very large number n of separate components which we will call, for brevity, molecules. We will assume that these molecules are not very different in respect to their structure, so that we will be able to use the approximate formulae derived in the previous section; as already indicated, the necessary conditions are always satisfied in actual physical problems. Let the systems G_1 and G_2 consist of n_1 and n_2 $(n_1 + n_2 = n)$ molecules. Let us also assume that the molecules forming the component G_1 possess the structure functions: $\omega_1(x), \cdots, \omega_{n_1}(x)$, the generating functions $\varphi_1(\alpha), \cdots, \varphi_{n_1}(\alpha)$, and the conjugate functions $u_1^{(\vartheta)}(x), \cdots, u_{n_1}^{(\vartheta)}(x)$, where ϑ is determined as the simple root of the equation:

$$-\frac{d \log \Phi(\alpha)}{d\alpha} = a.$$

Writing a_k and b_k for the mathematical expectation and the dispersion of the conjugate distribution $u_k^{(\vartheta)}(x)$, we have because of the formulae (35) and (36) (section 17, Chapter IV)

$$a_k = -\left(\frac{d \log \varphi_k(\alpha)}{d\alpha}\right)_{\alpha=\vartheta}, \qquad b_k = \left(\frac{d^2 \log \varphi_k(\alpha)}{d\alpha^2}\right)_{\alpha=\vartheta},$$

which we will write in the shorter form:

$$a_k = -\frac{d \log \varphi_k}{d\vartheta}, \qquad b_k = \frac{d^2 \log \varphi_k}{d\vartheta^2}.$$

Since, because of the assumed enumeration of the molecules

$$\Phi(\vartheta) = \prod_{k=1}^{n} \varphi_k(\vartheta), \ \Phi_1(\vartheta) = \prod_{k=1}^{n_1} \varphi_k(\vartheta), \ \Phi_2(\vartheta) = \prod_{k=n_1+1}^{n} \varphi_k(\vartheta),$$

we have

$$a = -\frac{d \log \Phi}{d\vartheta} = \sum_{k-1}^{n} a_k ,$$

$$A_1 = -\frac{d \log \Phi_1}{d\vartheta} = \sum_{k=1}^{n_1} a_k , \qquad A_2 = -\frac{d \log \Phi_2}{d\vartheta} = \sum_{k=n_1+1}^{n} a_k ,$$

$$A_1 + A_2 = a,$$

and also

$$B = \frac{d^2 \log \Phi}{d\vartheta^2} = \sum_{k=1}^{n} b_k ,$$

$$B_1 = \frac{d^2 \log \Phi_1}{d\vartheta^2} = \sum_{k=1}^{n_1} b_k , \qquad B_2 = \frac{d^2 \log \Phi_2}{d\vartheta^2} = \sum_{k=n_1+1}^{n} b_k ,$$

$$B_1 + B_2 = B.$$

From this relation it follows that the quantities a and B must be considered as infinitely large quantities of order n.

Let us assume now that the component G_1 represents a negligible small part of the entire system G, i.e. that n_1 is negligibly small in comparison with n (in particular, we can assume $n_1 = 1$, taking one single molecule as the component G_1). In our asymptotic formulae this condition is expressed by:

$$n_1 = 0(n) \qquad (n \to \infty).$$

Since we have agreed to consider all quantities a_k as well as all quantities b_k as being of the same order of magnitude we can conclude from the above given group of formulae that:

$$A_1 = 0(a), \qquad B_1 = 0(B) \qquad (n \to \infty),$$

and consequently

$$A_2 \sim A, \qquad B_2 \sim B \qquad (n \to \infty).$$

Keeping this in mind let us apply the approximate formulae

of the previous section to the expression (43). Because of the formula (42) section 19:

$$\Omega(a) = \frac{\Phi(\vartheta)e^{a\vartheta}}{(2\pi B)^{1/2}}\{1 + 0(1)\}. \tag{44}$$

To get the expression for $\Omega_2(a - E_1)$ we will use the formula (41) of the section 19. Obviously we must write $(a - E_1)$ instead of x, $\Phi_2(\vartheta)$ instead of $\Phi(\vartheta)$, B_2 instead of B, and A_2 instead of a. Since $A_2 = a - A_1$, the difference $x - a$ will be substituted by $A_1 - E_1$. In the remaining terms we can keep n (rather than substituting n_2 for it) because of the relation $n_2 \sim n$. Thus we find

$$\Omega_2(a - E_1) = \Phi_2(\vartheta)e^{\vartheta(a-E_1)}\left\{\frac{\exp\left[-\dfrac{(E_1 - A_1)^2}{2B_2}\right]}{(2\pi B_2)^{1/2}} + 0\left(\frac{1}{n}\right)\right\}$$

where $A_1 = 0(a) = 0(n)$. If we will consider only such values E_1 for which $E_1 - A_1 = 0(n^{1/2})$ the bracket on the right side of the above formula becomes:

$$\frac{1}{(2\pi B_2)^{1/2}}\{1 + 0(1)\},$$

and we will have:

$$\Omega_2(a - E_1) = \Phi_2(\vartheta)\frac{e^{\vartheta(a-E_1)}}{(2\pi B_2)^{1/2}}\{1 + 0(1)\}.$$

Comparing this with the formula (44), and noticing that $B_2 \sim B$ we obtain:

$$\frac{\Omega_2(a - E_1)}{\Omega(a)} = \frac{e^{-\vartheta E_1}}{\Phi_1(\vartheta)}\{1 + 0(1)\} \qquad (E_1 - A_1 = 0(n^{1/2})). \tag{45}$$

Thus we obtain for the distribution law of a small component, in its corresponding phase space, a very simple asymptotic formula (Boltzmann's law). The most important feature of this law is its exponential dependence on the energy of the small component in question, and the important role of the parameter

ϑ suggests that this parameter must have a simple physical interpretation.

Considering the energy E_1 of our small component, we can write for its probability density (according to the section 15, Chapter IV)

$$\frac{\Omega_1(x)\,\Omega_2(a - x)}{\Omega(a)}.$$

According to formula (45) we can put $| x - A_1 | = 0(n^{1/2})$ and write the above expression in the form:

$$(46) \qquad \frac{\Omega_1(x)e^{-\vartheta x}}{\Phi_1(\vartheta)}\,\{1 + 0(1)\}.$$

It must be noted that we have obtained for the approximate expression of the energy distribution of a small component the exact conjugate function of this component

$$U_1^{(\vartheta)}(x) = \frac{\Omega_1(x)e^{-\vartheta x}}{\Phi_1(\vartheta)}.$$

It is, of course, important that the parameter α assumes the value ϑ, i.e. satisfies the equation:

$$\frac{d \log \Phi}{d\alpha} + a = 0.$$

Thus, we see that the conjugate distribution law of a small component (in particular of a single molecule), taken for $\alpha = \vartheta$ permits a simple physical interpretation of the energy distribution of this component.

When G_1 is a single molecule $A_1 = a_1$ remains constant for increasing n, and the formula (46) applies uniformly when x varies within arbitrary constant limits.

The probability that the molecule will have an energy between g_1 and g_2 is given (for the i-th molecule) by the formula:

$$\int_{g_1}^{g_2} \frac{\omega_i(x)e^{-\vartheta x}}{\varphi_i(\vartheta)}\,dx\{1 + 0(1)\}.$$

Hence the mathematical expectation of the number of molecules with the energy between g_1 and g_2 is given by:

$$\sum_{i=1}^{n} \int_{g_1}^{g_2} \frac{\omega_i(x)e^{-\vartheta x}}{\varphi_i(\vartheta)}\, dx + 0(n)$$

and in the case when all molecules have identical structure (structure-function $\omega(x)$) we write:

$$n \int_{g_1}^{g_2} \frac{\omega(x)e^{-\vartheta x}}{\varphi(\vartheta)}\, dx + 0(n).$$

21. Mean values of the sum functions. In the present section we will consider the small component G_1 as being a separate molecule; since the enumeration of molecules is immaterial we will write $\omega_1(x)$ and $\varphi_1(\alpha)$ for the corresponding structure and generating functions of the selected molecule g_1.

Each phase function $f(x_1, x_2, \cdots)$, depending only on dynamical coordinates of the molecule g_1, can be interpreted as a function $f(P)$ of the point P in the phase space γ_1 of this molecule. Since the set of dynamic coordinates of the molecule g_1 has the probability density

$$\frac{\Omega^{(1)}(a - e_1)}{\Omega(a)}$$

where $\Omega^{(1)}(x)$ is the structure function of the complementary system $G - g_1$, and e_1 is the energy of the molecule g_1, the mean value of the function f is given by:

$$\bar{f} = \int_{\gamma_1} f(P) \frac{\Omega^{(1)}(a - e_1)}{\Omega(a)}\, dv_1$$

where dv_1 stands for the volume element in the phase space γ_1 of the molecule g_1, and it is assumed that the above integral converges absolutely.

Using the asymptotic formulae derived above, we can obtain the approximate expression of this integral, and estimate the corresponding error. To do this we break up the space γ_1 into two parts: γ_1' being the set of points in the space γ_1 for which

$e_1 < \log^2(n)$, and γ_1'' being the set of all other points. We put:

$$\int_{\gamma'_1} f(P) \frac{\Omega^{(1)}(a - e_1)}{\Omega(a)} dv_1 = I',$$

$$\int_{\gamma''_1} f(P) \frac{\Omega^{(1)}(a - e_1)}{\Omega(a)} dv_1 = I'',$$

so that

$$\bar{f} = I' + I''.$$

In order to guarantee the convergence of our integral, we will assume that, for large energy values e_1, the absolute value of $f(P)$ increases not faster than a certain power of this energy; i.e. that:

$$f(P) = 0(e_1^k) \qquad (e_1 \to \infty).$$

Next, let us evaluate the integral I''. Putting, as usual,

$$-\frac{d \log \varphi_1}{d\vartheta} = a_1, \qquad \frac{d^2 \log \varphi_1}{d\vartheta^2} = b_1$$

and writing $\Phi^{(1)}(\alpha)$ for the generating function of the system $G - g_1$, we can write (according to the formula (41) section 19):

$$\Omega^{(1)}(a - e_1) = \Phi^{(1)}(\vartheta)e^{\vartheta(a-e_1)}\left\{\frac{\exp\left[-\dfrac{(e_1 - a_1)^2}{2(B - b_1)}\right]}{[2\pi(B - b_1)]^{1/2}} + 0\left(\frac{1}{n}\right)\right\}$$

$$< \Phi^{(1)}(\vartheta)e^{\vartheta(a-e_1)}\left\{\frac{1}{[2\pi(B - b_1)]^{1/2}} + 0\left(\frac{1}{n}\right)\right\},$$

from which follows, for sufficiently large n, that

$$\Omega^{(1)}(a - e_1) < 2\Phi^{(1)}(\vartheta) \frac{e^{\vartheta(a-e_1)}}{(2\pi B)^{1/2}}.$$

On the other hand the formula (42) section 19 gives us, for sufficiently large n,

$$\Omega(a) > \frac{1}{2} \Phi(\vartheta) \frac{e^{\vartheta a}}{(2\pi B)^{1/2}}$$

which leads to:

$$\frac{\Omega^{(1)}(a - e_1)}{\Omega(a)} < \frac{4e^{-\vartheta e_1}}{\varphi_1(\vartheta)}$$

and consequently

$$| I'' | \leq \int_{\gamma''_1} | f(P) | \frac{\Omega^{(1)}(a - e_1)}{\Omega(a)} \, dv_1 < C \int_{\gamma''_1} \frac{e_1^k e^{-\vartheta e_1}}{\varphi_1(\vartheta)} \, dv_1 ,$$

where C is a positive constant. Because of the general formula (18) section 8, Chapter II

$$\int_{\gamma''_1} e_1^k e^{-\vartheta e_1} \, dv_1 = \int_{\log^2 n}^{\infty} x^k \omega_1(x) e^{-\vartheta x} \, dx,$$

we obtain

$$| I'' | < \frac{C}{\varphi_1(\vartheta)} \int_{\log^2 n}^{\infty} x^k \omega_1(x) e^{-\vartheta x} \, dx$$

$$< \frac{C}{\varphi_1(\vartheta)} \exp\left\{ - \frac{\vartheta}{2} \log^2 n \right\} \int_{\log^2 n}^{\infty} x^k \omega_1(x) e^{-\vartheta x/2} \, dx.$$

Since finally, the last integral in the above expression tends to zero for $n \to \infty$, we obtain, for sufficiently large n:

$$(47) \qquad | I'' | < \exp\left\{ - \frac{\vartheta}{2} \log^2 n \right\} < \frac{1}{n}.$$

Let us now evaluate the integral I'. Because of the formula (41) section 19, we have, for $e_1 < a_1 + \log^2 (n)$:

$$\Omega^{(1)}(a - e_1)$$

$$= \Phi^{(1)}(\vartheta) e^{\vartheta (a - e_1)} \left\{ \frac{\exp\left[- \dfrac{(e_1 - a_1)^2}{2(B - b_1)} \right]}{[2\pi(B - b_1)]^{1/2}} + 0\left(\frac{1 + | e_1 - a_1 |}{n^{3/2}} \right) \right\}$$

$$= \Phi^{(1)}(\vartheta) e^{\vartheta (a - e_1)} \left\{ 1 + 0\left(\frac{1 + (e_1 - a_1)^2}{n} \right) \right\} \frac{1}{(2\pi B)^{1/2}}$$

whereas the formula (42) section 19

$$\Omega(a) = \Phi(\vartheta)e^{\vartheta a}\frac{1}{(2\pi B)^{1/2}}\left\{1 + 0\left(\frac{1}{n}\right)\right\}.$$

It follows

$$\frac{\Omega^{(1)}(a - e_1)}{\Omega(a)} = \frac{e^{-\vartheta e_1}}{\varphi_1(\vartheta)}\left\{1 + 0\left(\frac{1 + (e_1 - a_1)^2}{n}\right)\right\}$$

and consequently

$$I' = \int_{\gamma'_1} f(P)\frac{e^{-\vartheta e_1}}{\varphi_1(\vartheta)}\left\{1 + 0\left(\frac{1 + (e_1 - a_1)^2}{n}\right)\right\} dv_1$$

$$= \int_{\gamma'_1} f(P)\frac{e^{-\vartheta e_1}}{\varphi_1(\vartheta)} dv_1 + 0\left(\frac{1}{n}\right).$$

The integration over the space γ_1' can be extended over γ_1 without difficulty since as we have seen above

$$\int_{\gamma''_1} f(P)\frac{e^{-\vartheta e_1}}{\varphi_1(\vartheta)} dv_1 = 0\left(\frac{1}{n}\right).$$

Thus taking into account (47), we obtain:

$$\bar{f} = \int_{\gamma_1} f(P)\frac{e^{-\vartheta e_1}}{\varphi_1(\vartheta)} dv_1 + 0\left(\frac{1}{n}\right). \tag{48}$$

In particular, when $f(P) = \chi(e_1)$ is a function of the energy of the selected molecule we obtain, because of the formula (19) section 8, Chapter II,

$$\begin{aligned}\bar{\chi} &= \int \chi(x)\frac{e^{-\vartheta x}}{\varphi_1(\vartheta)}\omega_1(x)\,dx + 0\left(\frac{1}{n}\right)\\ &= \int \chi(x)u_1^{(\vartheta)}(x)\,dx + 0\left(\frac{1}{n}\right).\end{aligned} \tag{49}$$

Thus, in particular,

$$\bar{e}_1 = \int x u_1^{(\vartheta)}(x)\,dx + 0\left(\frac{1}{n}\right) = a_1 + 0\left(\frac{1}{n}\right), \tag{50}$$

and

$$\overline{(e_1 - a_1)^2} = \int (x - a_1)^2 u_1^{(\vartheta)}(x)\, dx + 0\left(\frac{1}{n}\right) = b_1 + 0\left(\frac{1}{n}\right),$$

etc. This formula emphasizes the role of the conjugate function $u_1^{(\vartheta)}(x)$ as the approximate distribution law for the energy of a molecule.

Most of the phase functions which we encounter in statistical mechanics have a very special form. They are almost always the sums of functions each depending only on the dynamic coordinates of only one molecule. Such phase functions we will call sum functions. Thus if a system G is formed by the molecules $g_1, g_2, \cdots, g_n$ corresponding to the phase spaces γ_1, $\gamma_2, \cdots, \gamma_n$, the sum function can be written as:

$$f(P) = \sum_{i=1}^{n} f_i(P_i),$$

where P_i is some point in the space $\gamma_i (i = 1, 2, \cdots, n)$. Since the mathematical expectation of a sum is always equal to the sum of the mathematical expectations of its terms, we obtain (using formula (48)) the following approximate expression for the phase average for such a sum function:

$$\bar{f} = \sum_{i=1}^{n} \bar{f}_i = \sum_{i=1}^{n} \int_{\gamma_i} f_i(P_i) \frac{e^{-\vartheta e_i}}{\varphi_i(\vartheta)}\, dv_i + 0(1)$$

(it goes without saying that the functions $f_i(P_i)$ must satisfy the general assumptions used in the derivation of the formula (48)).

Example 1. *The number of molecules with energy within certain limits.* Let $0 \leq \alpha < \beta \leq +\infty$ and:

$$f_i(P_i) = \begin{cases} 1, & \text{if} \quad \alpha < e_i < \beta, \\ 0 & \text{in all other cases.} \end{cases}$$

The sum function

$$f(P) = \sum_{i=1}^{n} f_i(P_i)$$

represents, apparently, the number n_α^β of the molecules of the given system with the energy between α and β. According to (49):

$$\overline{n_\alpha^\beta} = \bar{f} = \sum_{i=1}^{n} \int_\alpha^\beta \omega_i(x) \frac{e^{-\vartheta x}}{\varphi_i(\vartheta)} dx + 0(1)$$

$$= \sum_{i=1}^{n} \int_\alpha^\beta u_i^{(\vartheta)}(x)\, dx + 0(1).$$

In particular when all molecules are identical we have:

$$\frac{\overline{n_\alpha^\beta}}{n} = \int_\alpha^\beta u^{(\vartheta)}(x)\, dx + 0\left(\frac{1}{n}\right).$$

Example 2. *The energy of a large component.* Let G_1 be the component of the system G consisting of the molecules g_1, g_2, $\cdots$, g_{n_1}. Let E_1 be the energy of this component and

$$f_i(P_i) = \begin{cases} e_i & (1 \le i \le n_1) \\ 0 & (n_1 < i \le n). \end{cases}$$

It is apparent that

$$E_1 = \sum_{i=1}^{n} f_i(P_i)$$

so that formula (50) gives us:

$$(51) \qquad \bar{E}_1 = \sum_{i=1}^{n_1} a_i + 0(1) = -\left(\frac{d \log \Phi_1(\alpha)}{d\alpha}\right)_{\alpha=\vartheta} + 0(1).$$

Let us note here that we cannot use the same method for the evaluation of the dispersion of E_1, since the energies of different molecules forming the component G_1 are not independent so that the dispersion of their sum is different from the sum of their dispersions. This considerably more complicated question will be discussed later (Chapter VIII).

22. ***Energy distribution law of a large component.*** The derivation of the asymptotic distribution law for the sum function considered as a random quantity is a rather difficult problem and will be considered in detail in one of the following chapters (Chapter VIII). However, in the most important case of the function E_1 considered in the previous section, the problem can be solved comparatively simply.

Let G_2 be the component complementary to G_1 , and let E_2 be its energy. We also put $n_2 = n - n_1$ and consider n_1 and n_2 as being infinitely large quantities of the order n. In general we will use the indices 1 and 2 to denote the quantities pertaining to G_1 and G_2 ; in particular we put:

$$A_1 = \sum_{i=1}^{n_1} a_i , \qquad A_2 = a - A_1 = \sum_{i=n_1+1}^{n} a_i ,$$

$$B_1 = \sum_{i=1}^{n_1} b_i , \qquad B_2 = B - B_1 = \sum_{i=n_1+1}^{n} b_i .$$

According to the formula (27) section 15, Chapter IV the probability density of E_1 is given by

$$\frac{\Omega_1(x)\,\Omega_2(a-x)}{\Omega(a)}.$$

According to (41) and (42) of section 19:

$$\Omega_1(x) \quad = \Phi_1(\vartheta)e^{\vartheta x}\left\{\frac{\exp\left[-\dfrac{(x-A_1)^2}{2B_1}\right]}{(2\pi B_1)^{1/2}} + 0\left(\frac{1}{n}\right)\right\},$$

$$\Omega_2(a-x) = \Phi_2(\vartheta)e^{\vartheta(a-x)}\left\{\frac{\exp\left[-\dfrac{(x-A_1)^2}{2B_2}\right]}{(2\pi B_2)^{1/2}} + 0\left(\frac{1}{n}\right)\right\},$$

$$\Omega(a) \quad = \Phi(\vartheta)e^{\vartheta a}\left\{\frac{1}{(2\pi B)^{1/2}} + 0\left(\frac{1}{n}\right)\right\}.$$

Writing, for brevity, $(B_1B_2)/B = B^*$ and remembering that, because of the law of composition for the generating functions $\Phi(\vartheta) = \Phi_1(\vartheta)\Phi_2(\vartheta)$, we get:

$$\frac{\Omega_1(x)\Omega_2(a-x)}{\Omega(a)}$$

$$(52) \qquad = \frac{1}{(2\pi B^*)^{1/2}} \exp\left\{-\frac{(x-A_1)^2}{2B_1} - \frac{(x-A_1)^2}{2B_2}\right\} + 0\left(\frac{1}{n}\right)$$

$$= \frac{1}{(2\pi B^*)^{1/2}} \exp\left\{-\frac{(x-A_1)^2}{2B^*}\right\} + 0\left(\frac{1}{n}\right).$$

Thus, for $n \to \infty$, the distribution of E_1 is given by the Gauss distribution function with the maximum at A_1 and the dispersion

$$B^* = \frac{B_1B_2}{B}.$$

If the energies of the molecules composing the component G_1 were mutually independent the dispersion of their sum would be

$$\sum_{i=1}^{n_1} b_i = B_1 .$$

Thus we see that the true dispersion $B^* < B_1$, which is quite understandable. In fact, since the sum of energies of all n molecules is fixed, the energies of individual molecules are correlated negatively, so that the dispersion of their sum is smaller than the sum of their dispersions.

23. *Example of monatomic ideal gas.* As an example of the application of the above discussed general methods we will consider now the simplest statistical system; a monatomic ideal gas. This served as the first example in the development of the basic ideas of statistical mechanics. Under the name "ideal monatomic gas" we will imagine a system G whose molecules $g_1, g_2, \cdots, g_n$ are simply material points. As usual, the total energy of the system is the sum of the energies of the individual molecules so that the molecules must not possess any mutual potential energy; as we have seen in section 8, Chapter II, this condition, which is unavoidable for the applicability of our methods, is actually never fulfilled in reality so that we have to consider it as only an approximation. We will assume that our gas (system G) is contained in a vessel with the finite volume V; this condition will be expressed formally by the introduction of a special term $U_i(x_i, y_i, z_i)$ representing the potential of the walls into the expression for the energy e_i of the molecule g_i (with the coordinates x_i, y_i, z_i). Since we assume that the system G is not subject to any outside forces we can write

$$e_i = \frac{m_i}{2}(\dot{x}_i^2 + \dot{y}_i^2 + \dot{z}_i^2) + U_i(x_i, y_i, z_i),$$

where m_i is a mass of the molecule g_i, whereas $\dot{x}_i, \dot{y}_i, \dot{z}_i$ are the components of its velocity. We will assume that the forces between the walls of the vessel and the molecules are different from zero only at very small distances from the wall. If we require that not a single molecule, no matter how fast it is moving, can penetrate through the walls of the vessel, we must assume that U_i is infinitely large outside the vessel. Inside the vessel we can assume U_i to be an arbitrary constant putting for simplicity $U_i = 0$. Of course, such a description of the function U_i ($U_i = 0$ inside, $U_i = +\infty$ outside) is only an approximate one; it would be more correct to assume that the function U_i is continuous and increases very rapidly when the molecule approaches the wall. However, we will use this idealized concept of U_i since it considerably simplifies the calculations without essentially affecting the results.

The Hamiltonian dynamic coordinates of the molecule g_i are represented by its three Cartesian coordinates and three components of its momentum,

$$p_i = \frac{\partial e_i}{\partial \dot{x}_i} = m_i\dot{x}_i, \qquad q_i = m_i\dot{y}_i, \qquad r_i = m_i\dot{z}_i.$$

Therefore,

$$e_i = \frac{1}{2m_i}(p_i^2 + q_i^2 + r_i^2) + U_i(x_i\,, y_i\,, z_i),$$

and consequently

$$E = \sum_{i=1}^{n} e_i = \sum_{i=1}^{n} \frac{1}{2m_i}(p_i^2 + q_i^2 + r_i^2) + \sum_{i=1}^{n} U_i(x_i\,, y_i\,, z_i)$$

where E is the total energy of the system G.

For the function $V(x)$ which expresses the part of the space Γ where $E < x$ we have the expression:

$$V(x) = \int_{E<x} dV = \int_{E<x} \prod_{i=1}^{n} dx_i\, dy_i\, dz_i\, dp_i\, dq_i\, dr_i\,.$$

Since outside of the vessel the potential energy, and consequently the total energy E, of the system G is infinitely large, the integration is carried out only inside the vessel. Since, on the other hand, the potential energy inside is equal to zero, we have:

$$V(x) = V^n \int_{\sum_{i=1}^{n} \frac{1}{2m_i}(p_i^2 + q_i^2 + r_i^2) < x} \prod_{i=1}^{n} dp_i\, dq_i\, dr_i\,.$$

The above integral represents the volume of an ellipsoid in the n-dimensional space with the semi-axes $(2m_i x)^{1/2}$ $(i = 1, 2, \cdots, n)$. This gives us

$$V(x) = V^n \frac{(2\pi)^{3n/2}}{\Gamma[(3n/2) + 1]} \left\{ \prod_{i=1}^{n} m_i^{3/2} \right\} x^{3n/2}.$$

Thus the structure function of the system G can be written as

$$\Omega(x) = \frac{dV(x)}{dx}$$

(53)

$$= V^n \frac{(2\pi)^{3n/2}}{\Gamma[(3n/2) + 1]} \left\{ \prod_{i=1}^{n} m_i^{3/2} \right\} \frac{3n}{2}\, x^{(3n/2)-1}.$$

In this elementary example the expression for the structure function is so simple that one does not require an approximate formula. However, for purposes of illustration we will construct the asymptotic formula for $\Omega(x)$ and will compare it with the exact expression (53). For the generating function $\Phi(\vartheta)$ of the system G we have:

$$\Phi(\vartheta) = \int \Omega(x)e^{-\vartheta x}\,dx$$

$$= V^n \frac{(2\pi)^{3n/2}}{\Gamma[(3n/2)+1]}\left\{\prod_{i=1}^{n} m_i^{3/2}\right\}\frac{3n}{2}\int_0^\infty x^{(3n/2)-1}e^{-\vartheta x}\,dx$$

(54)

$$= V^n(2\pi)^{3n/2}\left\{\prod_{i=1}^{n} m_i^{3/2}\right\}\vartheta^{-3n/2}.$$

If x is the total energy of the system G, the quantity ϑ is determined as the root of the equation:

$$-\frac{d\log\Phi}{d\vartheta} = \frac{3n}{2\vartheta} = x,$$

thus,

$$\vartheta = \frac{3n}{2x}, \tag{55}$$

and consequently:

$$\Phi(\vartheta) = V^n(2\pi)^{3n/2}\left\{\prod_{i=1}^{n} m_i^{3/2}\right\}\left(\frac{3n}{2}\right)^{-3n/2} x^{3n/2}.$$

From the formula (42) section 19, where

$$B = \frac{d^2\log\Phi}{d\vartheta^2} = \frac{3n}{2\vartheta^2} = \frac{2x^2}{3n},$$

we obtain the asymptotic expression:

$$\Omega(x) \approx \frac{\Phi(\vartheta)e^{\vartheta x}}{[2\pi(d^2\log\Phi/d\vartheta^2)]^{1/2}}$$

(56)

$$= V^n \frac{(2\pi)^{-3n/2}}{[(3n/2)]^{3n/2}e^{-3n/2}[2\pi\cdot(3n/2)]^{1/2}}\left\{\prod_{i=1}^{n} m_i^{3/2}\right\}\frac{3n}{2}x^{(3n/2)-1}.$$

Comparing the approximate expression (56) with the exact expression (53) we see that in our single case the method leads to the substitution of the quantity $\Gamma[(3n/2)+1]$ in the formula (53) by its asymptotic expression given by Sterling formula.

We will be satisfied for the time being with considering only this simplest system as an illustration of our general method; the complete theory of the monatomic ideal gas will be discussed in the next chapter.

24. The theorem of equipartition of energy. We have seen (section 20) that the conjugate distribution function

$$u_i^{(\vartheta)}(x) = \frac{\omega_i(x)e^{-\vartheta x}}{\varphi_i(\vartheta)}$$

for a small component (in particular for a molecule) of a given system G represents an approximate expression of the energy distribution of this component. In the case of the monatomic ideal gas we have

$$\omega_i(x) = V\,\frac{(2\pi m_i)^{3/2}}{\Gamma(3/2)}\,x^{1/2} = 2\pi V(2m_i)^{3/2}x^{1/2},$$

$$\varphi_i(\vartheta) = (2\pi m_i)^{3/2}V\vartheta^{-3/2}$$

(this formula could be derived in a similar way as the formulae (53) and (54) of the section 23 or can be obtained directly as the special case of these formulae for $n = 1$). Thus:

$$u_i^{(\vartheta)}(x) = \frac{2}{\pi^{1/2}}\,\vartheta^{3/2}x^{1/2}e^{-\vartheta x}.$$

In particular, for the mean value of the energy e_i of the molecule g_i we have an approximate expression:

$$\bar{e}_i \approx \int x u_i^{(\vartheta)}(x)\,dx = -\frac{d\log\varphi_i}{d\vartheta} = \frac{3}{2\vartheta}. \tag{57}$$

In the case of an ideal monatomic gas we have considered all molecules as possessing identical structure (i.e. the identical expression for energy in terms of the dynamic coordinates), although their masses could be different. We see now that the mean value of energy as well as its distribution law is the same for all molecules independent of their masses. Thus, in the mixture of two gases—a heavy one and a light one—the mean energy of a molecule is the same for both components. Furthermore this mean energy value and its distribution is independent of the volume of the vessel, being a universal function of the parameter ϑ. This result, pertaining to the mean energy of a molecule, represents a special case of a general theorem of statistical mechanics known as "the law of equipartition of energy among the degrees of freedom". The importance of this theorem lies in the fact that in many cases it permits us to find the mean energy of one or the other component of the system without almost any calculation. We will now give the general proof of this theorem, omitting however some possible generalizations.

Consider a system G whose component G_i possesses t degrees of freedom and is described by the Hamiltonian variables $q_1, \cdots, q_t, p_1, \cdots, p_t$. Let us assume that the total energy E_i of the system G_i is its kinetic energy. This can be written generally as a quadratic form in $p_1, p_2, \cdots, p_t$, with coefficients which may depend on $q_1, q_2, \cdots, q_t$. Let us denote it by $H(q_j, p_k)$. Writing, as usual, $V_i(x)$ for the volume of that part of the phase space occupied by the component G_i where $E_i < x$, we have

$$V_i(x) = \int_{H(q_j, p_k) < x} dq_1 \cdots dq_t \, dp_1 \cdots dp_t$$

$$= \int dq_1 \cdots dq_t \int_{H(q_j, p_k) < x} dp_1 \cdots dp_t ,$$

where the inner integral for the fixed values of $q_1, q_2, \cdots, q_t$, is evaluated within an ellipsoid in the t-dimensional space representing its volume. It is clear that this volume is proportional to $x^{t/2}$ and that the coefficient depends on $q_1, q_2, \cdots, q_t$. Thus:

$$V_i(x) = x^{t/2} \int \psi(q_1, \cdots, q_t) \, dq_1 \cdots dq_t = c_1 x^{t/2},$$

where c_1 (as well as c_2 and c_3 which will be introduced later) is a positive constant. We get from this

$$\Omega_1(x) = \frac{dV_i}{dx} = c_2 x^{(t/2)-1} \qquad (x > 0),$$

$$\Phi_i(\vartheta) = c_2 \int_0^\infty x^{(t/2)-1} e^{-\vartheta x}\, dx = c_3 \vartheta^{-t/2},$$

and consequently

$$\overline{E}_i \approx -\frac{d \log \Phi_i}{d\vartheta} = \frac{t}{2\vartheta}. \tag{58}$$

This relation expresses the theorem which we intended to prove, in fact it shows that the mean energy of a component of the given system is proportional to the corresponding number of degrees of freedom, the coefficient of proportionality being the quantity $1/2\vartheta$. In particular, since for the molecule of monatomic ideal gas $t = 3$, the formula (57) derived earlier represents a particular case of the general formula (58).

It is interesting to notice that the theorem of equipartition of energy which was proved using the approximate expression for the mean energy actually holds for the exact expression (of course in this case we get a somewhat different coefficient of proportionality; the quantity ϑ arose from our approximate analysis and does not exist at all in the exact theory). In order to prove this we must notice that the Hamiltonian function $H(q_i, p_k)$ of the selected component, being a quadratic form in the variables p_k, must satisfy the Euler relation

$$H = \frac{1}{2} \sum_{k=1}^{t} p_k \frac{\partial H}{\partial p_k};$$

and therefore

$$\begin{aligned} \overline{E}_i &= \frac{1}{\Omega(a)} \int_{\Sigma_a} \frac{H\, d\Sigma}{\operatorname{grad} E} = \frac{1}{2\Omega(a)} \sum_{k=1}^{t} \int_{\Sigma_a} p_k \frac{\partial H}{\partial p_k} \frac{d\Sigma}{\operatorname{grad} E} \\ &= \frac{1}{2\Omega(a)} \sum_{k=1}^{t} \int_{\Sigma_a} p_k \frac{\partial E}{\partial p_k} \frac{d\Sigma}{\operatorname{grad} E}, \end{aligned} \tag{59}$$

since in the expression of the total energy E the variables p_k only enter in H. Thus:

$$\frac{\partial E}{\partial p_k} = \frac{\partial H}{\partial p_k} \qquad (1 \leq k \leq t).$$

For each of the surface integrals on the right hand side of (59) a volume integral can be substituted according to the Green's theorem. Since $(1/\text{grad } E)\ \partial E/\partial p_k$ is the cosine of the angle between the outward normal to the surface and the p_k-axis, we have

$$\int_{\Sigma_a} F \frac{\partial E}{\partial p_k} \frac{d\Sigma}{\text{grad } E} = \int_{V_a} \frac{\partial F}{\partial p_k} dV;$$

where $F(q_i\ ,\ p_k)$ is an arbitrary function of the dynamic variables of the system G. In particular:

$$\int_{\Sigma_a} p_k \frac{\partial E}{\partial p_k} \frac{d\Sigma}{\text{grad } E} = \int_{V_a} \frac{\partial p_k}{\partial p_k} dV = V(a)$$

i.e. each such integral is equal to the volume of that part of the phase space of the system G where $E < a$. Therefore the formula (59) gives us:

$$\overline{E}_i = \frac{tV(a)}{2\Omega(a)}, \tag{60}$$

proving the above statement.

Comparing the exact formula (60) with the formula (58) which (according to (50) section 21) is exact up to terms of order $1/n$ we obtain

$$\vartheta = \frac{\Omega(a)}{V(a)} + 0\left(\frac{1}{n}\right) = \frac{d \log V}{da} + 0\left(\frac{1}{n}\right), \tag{61}$$

this gives us the approximate expression of the parameter ϑ in terms of $V(a)$ i.e. the approximate solution of the equation

$$-\frac{d \log \Phi}{d\vartheta} = a,$$

which determines the quantity ϑ. The formula (61) plays an important role in some theoretical studies.

Example. Rotational energy of a molecule in diatomic ideal gas. We will imagine a molecule of a diatomic gas as a pair of material points connected by a rigid massless axis of infinitely small length. The position of such a system in space is determined by five parameters for which we will choose three Cartesian coordinates x, y, z of one of the points and two angular coordinates φ and ψ determining the direction of the axis. Writing p_x, p_y, p_z, p_φ, p_ψ for the corresponding momenta, m for the mass of the molecule, and A for the moment of inertia of the second point with respect to a center of rotation at the first we can express the total kinetic energy of the molecule as the sum of the translational energy:

$$e_t = \frac{1}{2m}(p_x^2 + p_y^2 + p_z^2)$$

and the rotational energy:

$$e_r = \frac{1}{2A}\left(p_\varphi^2 + \frac{p_\psi^2}{\sin^2\varphi}\right) \tag{62}$$

(we shall see however that a knowledge of this expression is unimportant for the solution of our problem).

The molecule in question represents a component of the gas (whose other components may have, however, an entirely different structure). On the basis of a general definition of the component section 8, Chapter II we can consider each of the two sets of dynamical variables (x, y, z, p_x, p_y, p_z) and $(\varphi, \psi, p_\varphi, p_\psi)$ as an individual component of our gas; these two components can be considered as the fictitious "carriers" of the energies e_t and e_r corresponding to three and two degrees of freedom. The determination of the mean values of any of these components can be done without calculation by using the theorem of equipartition of energy. This theorem leads immediately to our previous formula for the translational energy, whereas for the rotational energy it gives

$$\bar{e}_r = 2 \cdot \frac{1}{2}\frac{d \log V(a)}{da} = \frac{d \log V(a)}{da} \approx \frac{1}{\vartheta}.$$

Thus we see that the theorem of equipartition of energy among the degrees of freedom saves us many difficult calculations which we would have to carry out in the case of a more complicated statistical system.

For the sake of completeness let us find the approximate expression for the distribution of e_r. Since:

$$v_r(x) = \int d\varphi\, d\psi\, dp_\varphi\, dp_\psi$$

$$= \int_0^\pi d\varphi \int_0^{2\pi} d\psi \iint_{p_\varphi^2 + \frac{p_\psi^2}{\sin^2\varphi} < 2Ax} dp_\varphi\, dp_\psi = 8\pi^2 Ax$$

(here the inner integral represents the area of an ellipse with the semi-axes $(2Ax)^{1/2}$ and $(2Ax)^{1/2} \sin\varphi$) we find that the structure function of the "fictitious carrier" of the energy e_r is

$$\omega_r(x) = v_r'(x) = 8\pi^2 A$$

so that the generating function is

$$\varphi_r(\vartheta) = 8\pi^2 A \int_0^\infty e^{-\vartheta x}\, dx = \frac{8\pi^2 A}{\vartheta}.$$

This gives us, by the way, the familiar result:

$$\bar{e}_r \approx -\frac{d \log \varphi_r(\vartheta)}{d\vartheta} = \frac{1}{\vartheta}.$$

The conjugate function $u_r^{(\vartheta)}(x)$ is determined by the expression

$$u_r^{(\vartheta)}(x) = \frac{\omega_r(x)e^{-\vartheta x}}{\varphi_r(\vartheta)} = \vartheta e^{-\vartheta x}$$

which is the approximate expression for the probability density of the quantity e_r. From the above it follows that

$$\mathbf{P}(g_1 < e_r < g_2) \approx \vartheta \int_{g_1}^{g_2} e^{-\vartheta x}\, dx = e^{-\vartheta g_1} - e^{-\vartheta g_2}.$$

Thus, the rotational energy of a diatomic molecule is subject to an exponential distribution depending exclusively on the parameter ϑ.

25. A system in thermal equilibrium. Canonical distribution of Gibbs. In the previous discussions we have been considering G as an isolated system which is not subject to energy exchange with the surrounding medium; thus the total energy E was always considered as a constant. It is clear that in actual physical systems such an assumption can be only approximately true since any actual physical system, no matter how well isolated, nevertheless undergoes some kind of energy interaction with its surroundings.

Another possible idealization consists in considering the system G as being a component of another much larger system G^*. In this case the component can freely exchange energy with its surroundings, i.e., with the other parts of the system G^*, and the energy E of the system G must be considered as a random quantity varying with time. Its distribution law can be derived on the basis of the general formula which we have proved earlier. It is clear that the question as to which of the two idealizations is closer to reality must be decided on the basis of purely physical considerations in each particular case.

We have seen above, Chapter IV, section 13, that in the first case (completely isolated system) the fundamental distribution law for the system G can be obtained as follows: the point P in the phase space Γ of the system G, representing the state of this system is always located on the surface Σ_a and is distributed with the surface density:

$$\frac{1}{\Omega(a) \text{ grad } E} \cdot \tag{63}$$

In the second case (freely interacting system) we arrive at an entirely different fundamental law. Since G is now a small component of the system G^*, the point P is no longer bound to any surface of constant energy, but can move freely in the space Γ; according to the previously derived distribution law for the small component (section 20), the point P is distributed in the space Γ according to the probability density whose approximate form is

$$\frac{1}{\Phi(\vartheta^*)} e^{-\vartheta^* E} \tag{64}$$

(the quantity marked by asterisks pertains to the system G^*, so that the quantity ϑ^* is a root of the equation $-[d \log \Phi^*(\alpha)]/d\alpha = E^*$).

This latter idealized picture is usually referred to as a system in thermal equilibrium. A uniform constant temperature in this case is due to the free interaction between the system G and its surroundings.

According to Gibbs the fundamental distribution law (63) corresponding to the first idealized picture is called a *micro-canonical* distribution, whereas the law (64) corresponding to the second idealized picture is known as *canonical* distribution.

The fundamental difference between these two distribution laws lies in the fact that whereas (63) gives the distribution on the surface Σ_a, (64) establishes the distribution in the entire phase space Γ.

Let us now consider the canonical distribution (64) in greater detail. As we know (section 20), the distribution of energy E in the system G (considered as a small component of the system G^*) is given by the probability density

$$\frac{\Omega(x)e^{-\vartheta^* x}}{\Phi(\vartheta^*)}$$

so that:

$$\overline{E} = \frac{\int x\Omega(x)e^{-\vartheta^* x}\, dx}{\Phi(\vartheta^*)} = -\left(\frac{d \log \Phi(\alpha)}{d\alpha}\right)_{\alpha = \vartheta^*}.$$

This shows that the parameter ϑ^* plays the same role in the second picture as the parameter ϑ did in the first one, provided that instead of the fixed value $E = a$ we introduce in the canonical distribution the quantity $\bar{E}$ representing the mathematical expectation of E.

Assume now that the system G consists of two components G_1 and G_2 with the energies E_1 and E_2. In order to obtain the distribution law of the system G_1 in its phase space Γ_1, we must integrate the expression (64) over the entire phase space Γ_2. Since $E = E_1 + E_2$ and $\Phi(\vartheta^*) = \Phi_1(\vartheta^*)\Phi_2(\vartheta^*)$ and also (according to (32) section 16):

$$\int_{\Gamma_2} e^{-\vartheta^* E_2}\, dV_2 = \Phi_2(\vartheta^*)$$

this integration yields

$$\int_{\Gamma_2} \frac{e^{-\vartheta^* E}}{\Phi(\vartheta^*)}\, dV_2 = \frac{e^{-\vartheta^* E_1}}{\Phi(\vartheta^*)} \int_{\Gamma_2} e^{-\vartheta^* E_2}\, dV_2 = \frac{e^{-\vartheta^* E_1}}{\Phi(\vartheta^*)} \Phi_2(\vartheta^*)$$

(65)

$$= \frac{e^{-\vartheta^* E_1}}{\Phi_1(\vartheta^*)}$$

which represents the probability density governing the distrition of the component G_1 in its phase space. Thus, we see that any component of the canonically distributed system is also distributed canonically with the same value of the parameter ϑ^*.

We know that the formula (65) holds also for small components in the microcanonical distribution as represented by the fundamental law (63). On the other hand, for the large components the formula (65) does not hold since it leads to the energy distribution law

$$\frac{\Omega_1(E_1)e^{-\vartheta^* E_1}}{\Phi_1(\vartheta^*)}$$

which differs quite essentially from the formula (52) section 22 for the distribution of the large component of the microcanonical system.

The relation:

$$\frac{e^{-\vartheta^* E}}{\Phi(\vartheta^*)} = \frac{e^{-\vartheta^* E_1}}{\Phi_1(\vartheta^*)} \cdot \frac{e^{-\vartheta^* E_2}}{\Phi_2(\vartheta^*)}$$

shows us that in the case of the canonical distribution for a system consisting of several components the distribution laws for these components must be combined as if they were mutually independent groups of random quantities. On the other hand, in the case of a microcanonical distribution the various components of a given system are considered as mutually dependent due to the constancy of the total energy. That is why, in spite of the fact that for small components the distribution laws based on (63) and (64) are almost identical, such is not the case for the large components.

It is clear that the computations based on the law (64) are considerably simpler than those based on the law (63), since it is easier to operate with independent random quantities. Therefore, we must ask ourselves: to what extent can we use, as an approximation, the fundamental law (64) instead of (63) for the calculation of the mean values of the phase functions in the microcanonical distributions?

We have already mentioned that the majority of phase functions which one encounters in statistical mechanics are sum functions; i.e., they can be written in the form of a sum of terms each depending on the coordinates of one molecule only. Because of the above mentioned similarity in the distribution laws of small components, the mean value of each term can be found approximately using the formulae belonging to the canonical distribution (this was a basic idea in our method of approximation). But the mean value of the sum is always equal to the sum of the mean values for dependent as well as for independent quantities; thus in calculation of the mean values of the sum functions we can, as an approximation, use the canonical (64) instead of microcanonical distribution (63). As mentioned above this change constitutes our method of approximation.

However, if we are looking for the mean value of a function which is not a sum function (but, for example, the square of a sum function), the substitution of (63) by (64) would lead, generally speaking, to a completely erroneous result; the mean value of such a function in the microcanonical distribution is entirely different from its mean value in the canonical distribution. The trivial example of that kind is given by the dispersion of the total energy of the given system: this quantity, which is apparently equal to zero in the microcanonical distribution, has a definite positive value in the canonical distribution. In the Chapter VIII we will see further examples of this kind.

The relation between the distribution laws (63) and (64) is unfortunately not clarified sufficiently in the existing texts on statistical mechanics. Thus, one often says that one can base the statistical mechanics on either the ergodic theory leading to the formula (63) or the "hypothesis of canonical distribution" i.e. the distribution law (64) which is introduced as a postulate, and that in the final count both points of view lead to the same results.

We see, however, what is the actual situation. The basic laws (63) and (64) correspond to two entirely different idealized pictures. Looking for the rational foundation, instead of referring only to the practical successes we will see that the ergodic theory, or some kind of its equivalent, is necessary in both cases, since the law (64) for a system in thermal equilibrium can be based only on the law (63) for the isolated system. Finally the statement that both theories lead to the same result is correct *only* within certain limits (for mean values of the sum functions); beyond these limits the introduction of quantities taken from one of these idealized pictures into the other would lead to very serious mistakes.

CHAPTER VI

IDEAL MONATOMIC GAS

26. Velocity distribution. Maxwell's law. In the present chapter we will use the fundamental formulae of the theory of the ideal monatomic gas as derived in sections 23, 24 of the previous chapter and will employ the same notation for the quantities involved. Let us choose one of the molecules of such a gas with mass m_i, and consider one of its velocity components, for example, $\dot{x}_i = (1/m_i)p_{xi}$ where p_{xi} is the component of the momentum of the molecule in the x-direction. Our problem is to find the distribution of the quantity $\dot{x}_i$. For this purpose we introduce the function: $f(x_i, y_i, z_i, p_{xi}, p_{yi}, p_{zi})$ which is defined in six-dimentional space γ_i by the relations

$$f = \begin{cases} 1, & \text{if } \dot{x}_i = \dfrac{1}{m_i} p_{xi} < z, \\ 0 & \text{otherwise} \end{cases}$$

where z is an arbitrary real number. It is clear that the mean value of the function f represents the probability of the inequality $\dot{x}_i < z$, or, in other words, gives us the distribution law of the quantity $\dot{x}_i$.

But, according to the formula (48) section 21, chapter V:

$$\bar{f} = \int_{\gamma_i} f(P) \frac{e^{-\vartheta e_i}}{\varphi_i(\vartheta)} dv_i + 0\left(\frac{1}{n}\right). \tag{66}$$

On the other hand, we have (section 23, chapter V):

$$e_i = \frac{m_i}{2} (\dot{x}_i^2 + \dot{y}_i^2 + \dot{z}_i^2) + U_i(x_i, y_i, z_i)$$

(67)

$$\varphi_i(\vartheta) = V(2\pi m_i)^{3/2} \vartheta^{-3/2}.$$

As we know, the presence of the term U_i in the expression for

e_i permits us to calculate the integral on the right hand side of the formula (66) by integrating only within the limits of vessel containing the gas, however, within the vessel $U_i = 0$. Since, finally

$$dp_{x_i} = m_i d\dot{x}_i , \qquad dp_{y_i} = m_i d\dot{y}_i , \qquad dp_{z_i} = m_i d\dot{z}_i ,$$

we have:

$$\bar{f} \approx Vm_i^3 \iiint_{\dot{x}_i < z} \frac{\exp\left[-\frac{\vartheta m_i}{2}\right](\dot{x}_i^2 + \dot{y}_i^2 + \dot{z}_i^2)}{V(2\pi m_i)^{3/2}\vartheta^{-3/2}} \, d\dot{x}_i \, d\dot{y}_i \, d\dot{z}_i$$

$$= \left(\frac{m_i\vartheta}{2\pi}\right)^{3/2} \int_{\dot{x}_i < z} \exp\left[-\frac{\vartheta m_i}{2} \dot{x}_i^2\right] d\dot{x}_i$$

$$\int \exp\left[-\frac{\vartheta m_i}{2} \dot{y}_i^2\right] d\dot{y}_i \int \exp\left[-\frac{\vartheta m_i}{2} \dot{z}_i^2\right] d\dot{z}_i ,$$

or, since each of the two complete integrals is equal to $[(2\pi)/(\vartheta m_i)]^{1/2}$:

$$\bar{f} \approx \left(\frac{m_i\vartheta}{2\pi}\right)^{1/2} \int_{t<z} \exp\left[-\frac{\vartheta m_i}{2} t^2\right] dt.$$

Thus we see that the quantity $\dot{x}_i$ is approximately distributed according to the Gauss law with the center at zero and with the dispersion $2/(m_i\vartheta)$. In particular, it means that the distribution law of velocity components (known as Maxwell's law) depends on the mass of the molecule. In the case of the gas mixture the velocity distribution, in contrast to the energy distribution, is different for different molecules; the heavy molecules have smaller dispersion than the light ones which physically means that the former move slower than the latter (this is quite natural since their kinetic energies must be the same).

27. The gas pressure. We know from physics the important role played by pressure in the theory of gases. It is impossible to build the statistical theory of monatomic ideal gas without first defining this physical notion in terms of our mechanical picture.

Imagine within the vessel containing a given mass of gas a fixed thin plate with the area S. During the given time interval $(t, t + \Delta t)$ one side of this plate is subject, generally speaking, to a number of impacts from gas molecules. Each of these impacts communicates to the plate a certain impulse. The sum of the components of these impulses perpendicular to the plate determines the pressure of the gas on our plate, or to be more exact, on one of its sides. The mean value of this impulse referred to unit time and unit area is called the pressure (of a given point and in a given direction).

To make this definition more exact let us choose on our plate an arbitrary point P and its arbitrary neighborhood Δs. The sum of impulse-components perpendicular to the plate which fall within the time interval $(t, t + \Delta t)$ into the area Δs depends on the state of gas at the moment t and represents consequently a phase function of our system. The mean value of this phase function is a quantity depending on Δt and Δs only, so that dividing it by $\Delta t \Delta s$ and letting the time interval Δt and the area Δs approach zero we obtain in the limit the quantity which we will call the gas pressure at the point P and in the given direction (as we shall see later, and as it is natural to expect on the basis of the symmetry considerations, the gas pressure of the point P does not depend on the direction chosen).

Since we can choose the coordinate system arbitrarily, we can assume, without loss of generality, that the area Δs is perpendicular to the x-axis. Suppose a certain molecule has the velocity components $\dot{x}$, $\dot{y}$, $\dot{z}$ at the moment t. Where should this molecule be at this moment in order that it might strike the selected side of the plate Δs within the time interval $(t, t + \Delta t)$? It is apparent that it must be within a sloping cylinder with base Δs, height $|\dot{x}| \Delta t$. Furthermore, the axis of the cylinder must be parallel to the vector $(\dot{x}, \dot{y}, \dot{z})$, and it must be constructed on that side of the plate Δs which is subject to molecular impact. Let us assume, to be definite, that this side of the plate is facing the negative direction of x-axis; in this case apparently we must have $\dot{x} > 0$ since otherwise impact is impossible. (We are neglecting here the possibility of collisions between the molecules under consideration during the time interval Δt.)

Let $\Omega(x)$ be the structure function of the entire mass of gas, and $\Omega_{(i)}^{i}(x)$ be the structure function of the system which is obtained by removing the above selected molecules from the gas. We know (comp. 25) that for a total gas energy E the probability density for the selected molecule in its phase space is given by

$$q = \frac{\Omega^{(i)}(E - e_i)}{\Omega(E)}$$

where the energy of the selected molecule e_i is a function of its six dynamic coordinates as determined by the formula (67). This density q is also a function of the same kind. If the point (x_i, y_i, z_i) is outside of the vessel containing the gas, $U_i(x_i, y_i, z_i)$ becomes infinite and $q = 0$ since $\Omega^{(i)}(E - e_i) = 0$. Inside of the vessel $U_i(x_i, y_i, z_i) = 0$, and e_i together with $\Omega^{(i)}(E - e_i)$ are functions of the velocity coordinates $\dot{x}_i, \dot{y}_i, \dot{z}_i$ only. The same is true for the density q. This function is determined more exactly by the formula (53) of chapter V, according to which the quantity $\Omega(E)$ is given by the product of V^n and a constant (here $x = E$) whereas the quantity $\Omega^{(i)}(E - e_i)$ (for the points within the vessel) is equal to the product of V^{n-1} and a certain quantity depending on $\dot{x}_i, \dot{y}_i, \dot{z}_i$ only. Therefore:

$$q = \frac{1}{V}\chi(\dot{x}_i, \dot{y}_i, \dot{z}_i) \tag{68}$$

where the form of the function χ can be determined (this, however, is of no importance for us at the moment).

The probability that the selected molecule will strike the area Δs within the given time interval Δt and with the velocity components in the interval $\dot{x}$ to $\dot{x} + d\dot{x}$, $\dot{y}$ to $\dot{y} + d\dot{y}$, and

$\dot{z}$ to $\dot{z} + d\dot{z}$, can be obtained by integrating the quantity

$$\frac{1}{V}\chi(\dot{x}, \dot{y}, \dot{z})\, d\dot{x}\, d\dot{y}\, d\dot{z}\, dx\, dy\, dz$$

over the entire volume of the above described cylinder, i.e. will be given by

$$\frac{1}{V} \mid \dot{x} \mid \Delta t \Delta s \chi(\dot{x}, \dot{y}, \dot{z})\, d\dot{x}\, d\dot{y}\, d\dot{z}.$$

In the process of the impact our molecule communicates to the area Δs the impulse $m_i \mid \dot{x} \mid$ in the direction of the x axis (m_i being the mass of the molecule). Remembering that this impulse is zero if at the moment t our molecule is located outside of the above described cylinder, we obtain for the mathematical expectation of the mean impulse communicated by the selected molecule to the area Δs in the direction $0x$ during the time interval Δt the expression

$$\text{(69)} \qquad \frac{m_i \Delta s \Delta t}{V} \iiint \dot{x}^2 \chi(\dot{x}, \dot{y}, \dot{z})\, d\dot{x}\, d\dot{y}\, d\dot{z}$$

where the integration is performed only for the positive values of $\dot{x}$ (and all values of $\dot{y}$ and $\dot{z}$).

We have been speaking so far about the first phase of impact during which the velocity of the incident molecule drops from $\dot{x}$ to zero and it communicates its original impulse in the direction $0x$ to the plate. This will be followed by a second phase during which the plate communicates to the molecule the impulse in the opposite direction. We can calculate the mathematical expectation of this new impulse in the same way, with the only difference that the integral in the expression (69) must be taken along the negative axis $0x$.

In order to obtain the mathematical expectation of the total impulse communicated to the area Δs during the time Δt in the direction $0x$ we must add both results, thus obtaining an expression of the type (69) in which the integration is extended for all coordinates within the limits $(-\infty, +\infty)$.

The integral (69) can be evaluated approximately without detailed calculation, since the integral

$$(70) \qquad \frac{m_i}{2} \iiint (\dot{x}^2 + \dot{y}^2 + \dot{z}^2)\chi(\dot{x}, \dot{y}, \dot{z})\, d\dot{x}\, d\dot{y}\, d\dot{z}$$

represents the mean value of kinetic energy of the selected molecule,[1] which, according to formula (57) chapter V, is equal to $3/2\vartheta$. Because of the symmetry in the three velocity components, the integral in (69) is equal to one third of the integral in (70), thus being equal to $1/(m_i\vartheta)$.

Consequently the mathematical expectation of the impulse becomes approximately

$$\frac{\Delta s \Delta t}{V\vartheta}$$

or

$$\frac{1}{V\vartheta}$$

per unit time per unit area. (In particular we notice that it does not depend on the mass of the selected molecule.)

If the number of molecules is n, the mean value of the total impulse communicated to the plate (in the normal direction) per unit area per unit time is:

$$(71) \qquad p = \frac{n}{\vartheta V}$$

which is usually known as the pressure of the gas. We see that the gas pressure does not depend on the location or the orientation of the plate being only a function of the volume, the

[1]In fact, according to (68), this mean value is represented by

$$\frac{m_i}{2} \int (\dot{x}^2 + \dot{y}^2 + \dot{z}^2) q\, dx\, dy\, dz\, d\dot{x}\, d\dot{y}\, d\dot{z}$$

$$= \frac{m_i}{2} \iiint (\dot{x}^2 + \dot{y}^2 + \dot{z}^2)\chi\, d\dot{x}\, d\dot{y}\, d\dot{z}.$$

number of molecules and the total energy of the gas (since ϑ is function of the total energy determined by the formula (55) chapter V).

From (71) we obtain:

$$pV = n/\vartheta$$

which coincides with the well known formula of Clopeyron if we put $\vartheta = 1/kT$ where k is the Boltzmann's constant and T the absolute temperature. In the next section we will discuss the validity of such a physical interpretation, and will now remark only that the relation

$$n/V = p\vartheta$$

(obtainable from (71)), in which ϑ is considered as some universal function of temperature, leads to the well known Avogadro law: for equal temperature and pressure all gases contain an equal number of molecules per unit volume.

28. Physical interpretation of the parameter ϑ We have seen that the parameter ϑ plays a very important role in statistical systems; it enters in an essential manner into the expressions for any physical characteristics of the system.

On the other hand the notion of the temperature which plays, as is well known, the fundamental role in thermodynamics has not found as yet any interpretation in our mechanical theory; in our attempts to build a purely mechanical theory of the heat processes we must try to find an interpretation of this physical notion in terms of the notions used above.

This comparison justifies an attempt to connect the physical meaning of the parameter ϑ with the temperature T of the system, although, of course, it does not justify postulating a universal functional dependence between ϑ and T. We have seen in the previous section that in the case of a monatomic ideal gas one can give definite arguments in favor of such a postulate, and that in this case one can arrive at the relation

(72) $$\vartheta = \frac{1}{kT}$$

where k is Boltzmann's constant. However, even within the theory of the ideal monatomic gas it would be reasonable to explore the immediate consequences of the above postulate before accepting it. Thus we could substitute the expression (72) for ϑ in the distributions so far obtained (for example, in the section 26); we would see in this case that these distributions coincide exactly with those derived in physics on the basis of entirely different considerations.

Since the postulate (72) demands a careful check even in the simple theory of the ideal monatomic gas, it is clear that at the present moment we can tell hardly anything more definite about general dependence between ϑ and T in the case of more complicated physical systems. We will see, however, in the next chapter that the interpretation of the fundamental laws of thermodynamics on the basis of our general mechanical theory will lead quite naturally to a generalization of the postulate (72) for a very broad class of physical systems.

In the present chapter we will give only one additional argument in favor of a universal dependence between ϑ and the temperature of the system.

Suppose we have two physical systems with temperatures T_1 and T_2 which are characterized by the values ϑ_1 and ϑ_2 of our statistical parameter.

If we unite the two systems (T_1, ϑ_1) and (T_2, ϑ_2) into one system (T, ϑ) (i.e. if we assume that these two systems are interacting with one another), the finite temperature T of the composite system will lie between T_1 and T_2; in particular in the case T_1 is equal to T_2 we will have $T = T_1 = T_2$. If we assume that ϑ is a monotonic function of temperature its value must be intermediate between ϑ_1 and ϑ_2 for the composite system. It is easy to see that ϑ actually possesses this property.

In fact let $\Phi_1(\alpha)$, $\Phi_2(\alpha)$ and $\Phi(\alpha)$ be the generating functions of the two components and the composite system, and E_1, E_2 and E the corresponding total energies (so that $E = E_1 + E_2$ and $\Phi(\alpha) = \Phi_1(\alpha)\Phi_2(\alpha)$).

We know that:

$$E_1 = \left(-\frac{d \log \Phi_1}{d\alpha}\right)_{\alpha=\vartheta_1}, \quad E_2 = \left(-\frac{d \log \Phi_2}{d\alpha}\right)_{\alpha=\vartheta_2},$$

$$E = \left(-\frac{d \log \Phi}{d\alpha}\right)_{\alpha=\vartheta};$$

so that

$$(73) \qquad \left(\frac{d \log \Phi}{d\alpha}\right)_{\alpha=\vartheta} = \left(\frac{d \log \Phi_1}{d\alpha}\right)_{\alpha=\vartheta_1} + \left(\frac{d \log \Phi_2}{d\alpha}\right)_{\alpha=\vartheta_2}.$$

On the other hand, for arbitrary α, and, in particular, for $\alpha = \vartheta$

$$(74) \qquad \frac{d \log \Phi}{d\alpha} = \frac{d \log \Phi_1}{d\alpha} + \frac{d \log \Phi_2}{d\alpha}.$$

Let us assume for the moment that ϑ lies outside of the interval $(\vartheta_1, \vartheta_2)$, being, for example, larger than ϑ_1 and ϑ_2. Since according to the property 4 (chapter IV, section 16) the logarithmic derivatives of the leading functions increase with increasing value of the argument, we would get

$$\left(\frac{d \log \Phi_1}{d\alpha}\right)_{\alpha=\vartheta} > \left(\frac{d \log \Phi_1}{d\alpha}\right)_{\alpha=\vartheta_1},$$

$$\left(\frac{d \log \Phi_2}{d\alpha}\right)_{\alpha=\vartheta} > \left(\frac{d \log \Phi_2}{d\alpha}\right)_{\alpha=\vartheta_2};$$

thus for $\alpha = \vartheta$ the right hand side of the equation (74) is larger than the right hand side of the equation (73) whereas their left hand sides are equal to one another. This contradiction proves the incorrectness of the above assumption. This argument holds apparently for any physical system of the general type considered above.

29. Gas pressure in an arbitrary field of force. We will now return to the case of the ideal monatomic gas, but will assume that its molecules are subject to the action of an external field of force, and that the potential energy of each molecule is a function only of its position (i.e. independent of its velocity and of the position and velocities of other molecules).

In other words we will assume that the potential energy of each individual molecule can be written in the form

$$(75) \qquad e_i = \frac{m_i}{2}(\dot{x}_i^2 + \dot{y}_i^2 + \dot{z}_i^2) + \epsilon_i(x_i , y_i , z_i),$$

where $\epsilon_i(x_i , y_i , z_i)$ is the potential energy in the given field.

We will define the gas pressure p at a given point and in a given direction in the same way as was done in section 27. The distribution law for a selected molecule of our gas can be written (in the notation of the section 27) in the form:

$$(76) \qquad \frac{\Omega^{(i)}(E - e_i)}{\Omega(E)},$$

where e_i (the energy of the selected molecule) is a function of six dynamic coordinates of that molecule as determined by the formula (75), $\Omega^{(i)}(x)$ is the structure function of the system consisting of the entire mass of the gas minus the selected molecule. At a certain moment t the selected molecule is located within a certain cell of the phase space; in order that this molecule would hit the area Δs within the time interval) t, $t + \Delta t$) it is necessary and sufficient that this cell is a cylinder with the base Δs and the height $\dot{x}\Delta t(\dot{x} > 0)$. The impulse communicated by this molecule to the area Δs is $m_i\dot{x}$. We are neglecting here the effect of the collisions between the molecules as well as the possible deviations from rectlinear motion caused by the external field; in fact, these effects vanish when Δs and Δt approach to zero.

Thus, in order to calculate the value of the mean impulse communicated by the selected molecule to the area Δs in the direction $0x$ during the time interval Δt, we must integrate the product of $m_i\dot{x}$ and (76) over the six dimensional phase space of the selected molecule; the integration must be carried over the region in which $\dot{x} > 0$ and the limits of integration are determined by the condition that the selected molecule lies within the above described cylindrical volume. Since all dimensions of this cylinder are small quantities of the order of Δs and Δt, we can consider the values of the space coordinates entering into the integral as certain fixed quantities (for example, making them equal to the coordinates of a certain point P within the area Δs and towards which this area contracts in the limiting case). We can repeat here all the arguments of section 27 pertaining to the second phase of the impact, and come to the conclusion that in order to calculate the value of the total impulse we must extend the integration along the entire real axis $0\dot{x}$. Thus we can write for the mean value of the impulse

$$\frac{1}{\Omega(E)} \iiint m_i \dot{x} \cdot \dot{x} \Delta t \Delta s \Omega^{(i)}(E - e_i)\, dp_x\, dp_y\, dp_z$$

$$= \frac{m_i^4 \Delta t \Delta s}{\Omega(E)} \iiint \dot{x}^2 \Omega^{(i)}(E - e_i)\, d\dot{x}_i\, d\dot{y}_i\, d\dot{z}_i$$

or

$$\frac{m_i^4}{\Omega(E)} \iiint \dot{x}_i^2 \Omega^{(i)}(E - e_i)\, d\dot{x}_i\, d\dot{y}_i\, d\dot{z}_i$$

per unit area. As already mentioned, the integration must be extended through the entire three dimensional space $(\dot{x}_i, \dot{y}_i, \dot{z}_i)$, whereas the space coordinates x_i, y_i, z_i may be set equal to the coordinates x, y, z at the point P at which the gas pressure is being calculated. In order to get the pressure of the gas we have to take the sum of the above expressions for all n molecules of the gas. This gives:

$$p = \sum_{i=1}^{n} \frac{m_i^4}{\Omega(E)} \iiint \dot{x}_i^2 \Omega^{(i)}(E - e_i)\, d\dot{x}_i\, d\dot{y}_i\, d\dot{z}_i\,.$$

It is clear that in this more general case the pressure of the gas is different at different points since the above expression for p depends on the coordinates x, y, z of the point P.

In considering an approximate expression we will consider only the leading terms, leaving the reader the estimate of the corresponding errors (which can be done easily on the basis of the formula in chapter V).

To approximate the expression (76) we will use the formula (45) of chapter V. (Although this formula was established for values of e_i deviating from the mean value by an amount of the order of magnitude $0(n^{1/2})$, it is easy to verify that the part of the integral in the expression for p which corresponds to the large deviations is negligibly small as compared with the main part. Thus in the approximate evaluation of this integral we can either neglect this part or, what is more convenient, use for the integrated function the same expression as in the main part. This gives us:

$$p \approx \sum_{i=1}^{n} m_i^4 \iiint \dot{x}_i^2 \frac{e^{-\vartheta e_i}}{\varphi_i(\vartheta)} \, d\dot{x}_i \, d\dot{y}_i \, d\dot{z}_i$$

where

$$\varphi_i(\vartheta) = \int_{\gamma_i} e^{-\vartheta e_i} \, dv_i$$

is a generating function of the selected molecule.

We now introduce for the energy e_i its expression given by (75) remembering that in this expression for the coordinates x_i, y_i, z_i of the selected molecule must be substituted the coordinates x, y, z of the point P. Since

$$\varphi_i(\vartheta) = \int_{\gamma_i} e^{-\vartheta e_i} \, dv_i = \iiint \exp\left[-\vartheta \epsilon_i(x_i, y_i, z_i)\right] dx_i \, dy_i \, dz_i$$

$$\cdot \iiint \exp\left[-\frac{\vartheta m_i}{2}(\dot{x}_i^2 + \dot{y}_i^2 + \dot{z}_i^2)\right] dp_{xi} \, dp_{yi} \, dp_{zi}$$

(where $p_{xi} = m_i\dot{x}_i$, $p_{yi} = m_i\dot{y}_i$, $p_{zi} = m_i\dot{z}_i$) we obtain, after obvious simplifications:

$$p \approx \sum_{i=1}^{n} \frac{\exp\left[-\vartheta \epsilon_i(x, y, z)\right]}{\iiint \exp\left[-\vartheta \epsilon_i(x_i, y_i, z_i)\right] dx_i \, dy_i \, dz_i}$$

$$\cdot \frac{m_i \int_{-\infty}^{+\infty} \dot{x}_i^2 \exp\left[-\frac{\vartheta m_i}{2}\dot{x}_i^2\right] d\dot{x}_i}{\int_{-\infty}^{+\infty} \exp\left[-\frac{\vartheta m_i}{2}\dot{x}_i^2\right] d\dot{x}_i}$$

or since

$$\int_{-\infty}^{+\infty} u^2 \exp\left[-\frac{\vartheta m_i}{2}u^2\right] du = (2\pi)^{1/2}(\vartheta m_i)^{-3/2},$$

$$\int_{-\infty}^{+\infty} \exp\left[-\frac{\vartheta m_i}{2}u^2\right] du \quad = (2\pi)^{1/2}(\vartheta m_i)^{-1/2},$$

we get finally

$$(77) \qquad p \approx \frac{1}{\vartheta} \sum_{i=1}^{n} \frac{\exp\left[-\vartheta \epsilon_i(x, y, z)\right]}{\iiint \exp\left[-\vartheta \epsilon_i(x_i, y_i, z_i)\right] dx_i\, dy_i\, dz_i}.$$

In the particular case considered in section 27 ϵ_i is equal to zero inside of the vessel and to $+\infty$ outside of it, which gives us:

$$p \approx \frac{n}{\vartheta V}$$

where V is the volume of the vessel. This is exactly the expression which we obtained before.

We will consider as an example the behavior of a gas enclosed in a vessel and subjected to the action of gravitational force along the negative z direction. In this case the potential energy is

$$\epsilon_i(x, y, z) = m_i g z + U(x, y, z)$$

where $U(x, y, z) = 0$ inside the vessel and $= +\infty$ outside of it.

According to the formula (45) chapter V the distribution law of a single molecule is given by the approximate expression of the density:

$$(78) \qquad \frac{1}{\varphi_i(\vartheta)} \exp\left[-\frac{\vartheta m_i}{2}(\dot{x}_i^2 + \dot{y}_i^2 + \dot{z}_i^2) - \vartheta m_i g z_i\right]$$

inside the vessel (outside the vessel the density is, of course, equal to zero).

Assume for simplicity that all molecules have the same mass $m_i = m$. Let us select an arbitrary point $P_0(x_0, y_0, z_0)$ and imagine it to be surrounded by an elementary volume dv_0. The probability that some molecule will fall within this volume can be obtained by multiplying dv_0 by the expression (78) integrated over all three velocity coordinates from $-\infty$ to $+\infty$ (here, of course, $z_i = z_0$). This probability will, obviously, have the form

$$Ae^{-mgz_0\vartheta}\, dv_0$$

where A is a function of m and ϑ. The mean number of molecules which fall within the volume dv_0 is therefore:

$$nAe^{-mgz_0\vartheta}\, dv_0 .$$

At another point $P(x, y, z)$ within the volume dv this number is given by:

$$nAe^{-mgz\vartheta}\, dv.$$

If $dv = dv_0$ the ratio of these numbers (i.e., the density relative to P_0) is equal to

$$\exp\,[-mg\vartheta(z - z_0)].$$

If we put, as above, $\vartheta = 1/kT$, we obtain for the relative density of the gas the well known "barometric" formula

$$\exp\left[-\frac{mg}{kT}(z - z_0)\right]$$

which is usually derived in physics on the basis of entirely different considerations. It is clear that formula (77) (for $m_i = m$) leads us to an identical expression for the relative gas pressure.

CHAPTER VII

THE FOUNDATION OF THERMODYNAMICS

30. External parameters and the mean values of external forces. In the previous chapters we have often considered the case where the energy of the molecules forming the given physical system depends not only on the dynamic coordinates of these molecules, but also on the number of parameters characterizing the position or the state of external bodies acting on the system in question. Thus, for example, in the previous section the quantity g, characterizing the gravitational field, entered in a natural way into all the pertinent preceding formulae. In other cases such parameters can be represented, for example, by the coordinates of some attraction or repulsion-centers. In the future such parameters will be referred to as *external* parameters. Mathematically such an external parameter is characterized by the fact that it has the same form in the energy expressions for all molecules.

However, in all the above considerations, we have assumed that the values of the external parameters always remain constant; we will now concentrate on the cases when the external parameters change with time. We remark that the energies e_i of the individual molecules as well as the total energy $E = \Sigma e_i$ of the system are functions of the external parameters which we will in general denote by $\lambda_1, \cdots, \lambda_r$. A change of the external parameters (such as the change of the field of forces, the change of the position of the attraction—or repulsion—centers etc.) will result, generally speaking, in a change in the energy of the system; the point representing the system in its phase space will in this case execute a transition from one surface of constant energy to another. This change of energy is due to the work done by such external forces as change these parameters.

The quantity $\partial e_i/\partial\lambda_s$ will be called the generalized external force acting on the i-th molecule "in the direction" of the parameter λ_s. Similarly the quantity

$$X_s = \frac{\partial E}{\partial \lambda_s} = \sum_i \frac{\partial e_i}{\partial \lambda_s}$$

will be called the generalized external force in the direction of the parameter λ_s acting on the entire system. Apart from the parameters $\lambda_1, \cdots, \lambda_r$, the quantity X_s depends on all the dynamical coordinates of the given system, i.e., on the exact position of the representative point of the phase space. In particular, the quantity X_s may have entirely different values for two different points on the same surface of the constant energy. Thus from the point of view of our theory the quantity X_s is a phase function, or, in the terminology of the theory of probabilities, a random quantity. Thus it is natural to ask ourselves about the mean value $\overline{X_s}$ of this quantity on a given energy surface. According to the formula (48) chapter V we have:

$$\overline{\frac{\partial e_i}{\partial \lambda_s}} = \int_{\gamma_i} \frac{\partial e_i}{\partial \lambda_s} \frac{e^{-\vartheta e_i}}{\varphi_i(\vartheta)} dv_i + 0\left(\frac{1}{n}\right)$$

where γ_i, φ_i, and dv_i stand for the phase space, the generating function, and the element of volume in the phase space of the selected molecule, whereas the quantity λ_s is related to the total energy of the system E in the same way as in the previous cases.

But since

$$\varphi_i(\vartheta) = \int_{\gamma_i} e^{-\vartheta e_i} dv_i$$

we have

$$\int_{\gamma_i} \frac{\partial e_i}{\partial \lambda_s} \frac{e^{-\vartheta e_i}}{\varphi_i(\vartheta)} dv_i = -\frac{1}{\vartheta} \frac{\partial \log \varphi_i}{\partial \lambda_s}$$

and consequently

$$\overline{\frac{\partial e_i}{\partial \lambda_s}} \approx -\frac{1}{\vartheta} \frac{\partial \log \varphi_i}{\partial \lambda_s}$$

from which follows, approximately,

$$(79) \quad \overline{X}_s = \overline{\frac{\partial E}{\partial \lambda_s}} = \sum_i \overline{\frac{\partial e_i}{\partial \lambda_s}} = -\frac{1}{\vartheta} \sum_i \frac{\partial \log \varphi_i}{\partial \lambda_s} = -\frac{1}{\vartheta} \frac{\partial \log \Phi}{\partial \lambda_s}$$

where $\Phi = \Phi(\vartheta)$ is the leading function of the system, depending, apart from ϑ, on all external parameters. The above formula represents an extremely simple expression for the desired mean value.

The element of work done by the external forces when the external parameters change by $d\lambda_1, \cdots, d\lambda_r$, will be defined as usual, by the expression:

$$\delta A = \sum_{s=1}^{r} X_s \, d\lambda_s .$$

Similar to the quantities X_s, the quantity δA is a certain phase function (a random quantity). Calculating the mean value of δA for a given value E of the total energy of the system, we obtain:

$$(80) \qquad \overline{\delta A} = \sum_{s=1}^{r} \overline{X}_s \, d\lambda_s = -\frac{1}{\vartheta} \sum_{s=1}^{r} \frac{\partial \log \Phi}{\partial \lambda_s} d\lambda_s .$$

It is important to remember that the sum in the right hand side of the above expression does not represent the total differential of the function Φ, since, apart of the external parameters, this function also depends on the parameter ϑ.

31. The volume of the gas as an external parameter. One of the most important parameters encountered in the study of gases is the volume of the vessel containing the gas. This volume can usually be considered as a function of only one or a few external parameters. Let us consider the simple case of a gas enclosed in a cylindrical vessel with a movable piston; here the shape of the vessel is completely determined by the volume V. Thus, if the only forces acting on the gas are due to the reaction of the walls, the function $U(x_i, y_i, z_i)$, representing the potential energy of the molecule at the point (x_i, y_i, z_i), is completely determined by the quantity V; this justifies consideration of the quantity V as the external parameter of our system.

Let us find the expression for the mean value of the generalized force acting along this parameter. If we consider the case of an ideal gas, and use the formula (54), we obtain

$$\Phi(\vartheta) = (2\pi)^{3n/2}\left\{\prod_{i=1}^{n} m_i^{3/2}\right\}\vartheta^{-3n/2}V^n.$$

According to the formula (79) of the previous section, the mean value of the force becomes

$$\overline{X}_V = -\frac{1}{\vartheta}\frac{\partial \log \Phi}{\partial V} = -\frac{1}{\vartheta}\frac{n}{V}.$$

The above expression is equal in magnitude and opposite in sign to the expression for the pressure of the gas derived in the previous chapter. Thus we can consider the gas pressure as the mean value of the force with which this gas acts on the external bodies "in the direction" of the parameter V. In particular, the mean value of the elementary work done by the gas when its volume changes by dV, can be written as:

$$-\overline{\delta A} = p\,dV.$$

32. The second law of thermodynamics. The science of thermodynamics is based essentially on its two fundamental laws; thus, every theory pretending to represent the foundation of thermodynamics must prove that these two fundamental laws can be derived from its basic principles. Once this is done, the entire system of thermodynamic theory can be developed logically as a consequence of the two laws.

The first fundamental law of thermodynamics is the law of conservation of energy; it is clear that any mechanical foundation of the theory of heat includes this law quite automatically, since the law of conservation of energy represents the first integral of the equations of motion.

We encounter an entirely different situation in the case of the second fundamental law which, in the frame of the mechanical theory, presents a theorem subject to mathematical proof.

In the customary (non statistical) treatment of thermodynamics the state of a physical system is usually characterized by a set of external parameters and by the temperature of the system. We have seen before that there are many reasons to assume that the parameter ϑ is directly connected with the notion of the temperature of the system, and that, on the other hand a knowledge of ϑ is equivalent to a knowledge of the total energy E of the system. Thus, in the frame of our theory, the state of the system is completely determined if, in addition to the value of the external parameters, we also know the surface of constant energy which contains the representative point of our system. Thus, in the classical treatment, we do not distinguish between the states of the system represented by different points on the same energy surface. Because of this fact we will agree to use the term "a thermodynamic function" for any quantity which is completely determined by parameters $\vartheta, \lambda_1, \cdots, \lambda_r$. We have already encountered such "thermodynamic functions" in many places in the previous discussion. As an example, we can mention the generating function $\Phi(\vartheta)$, the total energy of the system $E = -(\partial \log \Phi)/\partial\vartheta$ (we use here the partial derivative to underline the fact that the values of parameters $\lambda_1, \cdots, \lambda_r$ must be kept constant) and finally the mean values of the acting forces

$$\overline{X}_s = \frac{1}{\vartheta} \frac{\partial \log \Phi(\vartheta)}{\partial \lambda_s} \qquad (1 \leq s \leq r).$$

It is clear that any thermodynamic function is at the same time a phase function subject to the condition of having a constant value on the surface of constant energy. Conversely, any phase function which is constant on a surface of given energy can be considered as a thermodynamic function.

Consider the transition of a system from one of its states $Z_1(\vartheta, \lambda_1, \cdots, \lambda_r)$, (determined in the above described classical thermodynamical sense) into another "immediately adjoining" state $Z_2(\vartheta + d\vartheta, \lambda_1 + d\lambda_1, \cdots, \lambda_r + d\lambda_r)$. The work done by the system in this transition is given by:

$$-\delta A = \sum_{s=1}^{r} - X_s \, d\lambda_s = - \sum_{s=1}^{r} \frac{\partial E}{\partial \lambda_s} d\lambda_s .$$

It must be remembered here that the generalized forces X_s are not phase, but rather thermodynamic, functions depending on the entire set of the phase coordinates (including, of course, the external parameters). These forces (and consequently also the work $-\delta A$) are not determined by the knowledge of the states Z_1 and Z_2, since a given thermodynamical state of a system corresponds to an entire continuum of individual states in the sense of statistical mechanics (the entire energy surface in the phase space of the system). From the point of view of statistical mechanics the work $-\delta A$ is not determined uniquely by the original and final thermodynamic characteristics of the system; in fact, it can have entirely different values depending on the exact positions of the representative point of the system in its phase space. This indicates that the quantity $-\delta A$ cannot be identified with the elementary work in the sense of physical thermodynamics.

As we have often seen above, such a situation is typical in our theory; in any attempt to build a bridge between statistical mechanics and any physical theory the role of physical quantities is played by not phase functions themselves but rather by their mean values taken over a given thermodynamical state of a system.

In our case the equivalent of the elementary work as considered in classical thermodynamics, is not the phase function $-\delta A$ but its mean value on a given surface of constant energy:

$$-\overline{\delta A} = - \sum_{s=1}^{r} \overline{X}_s \, d\lambda_s = \frac{1}{\vartheta} \sum_{s=1}^{r} \frac{\partial \log \Phi}{\partial \lambda_s} d\lambda_s .$$

Writing dE for the total energy change of the system in its transition from the state Z_1 into the state Z_2 we have

$$dE = \frac{\partial E}{\partial \vartheta} d\vartheta + \sum_{s=1}^{r} \frac{\partial E}{\partial \lambda_s} d\lambda_s .$$

The above expression is determined uniquely by the states Z_1 and Z_2 since E is a thermodynamical function with a well defined value for each thermodynamically described state. But

$$E = -\frac{\partial \log \Phi}{\partial \vartheta}$$

and consequently

$$-\overline{\delta A} = \frac{1}{\vartheta} \sum_{s=1}^{r} \frac{\partial \log \Phi}{\partial \lambda_s} d\lambda_s = \frac{1}{\vartheta}\left(d \log \Phi - \frac{\partial \log \Phi}{\partial \vartheta} d\vartheta\right)$$

$$= \frac{1}{\vartheta} [d \log \Phi + E d\vartheta].$$

Therefore,

$$\vartheta(dE - \overline{\delta A}) = \vartheta \frac{\partial E}{\partial \vartheta} d\vartheta + \vartheta \sum_{s=1}^{r} \frac{\partial E}{\partial \lambda_s} d\lambda_s + d \log \Phi + E d\vartheta$$

$$= \frac{\partial(\vartheta E)}{\partial \vartheta} d\vartheta + \sum_{s=1}^{r} \frac{\partial(\vartheta E)}{\partial \lambda_s} d\lambda_s + d \log \Phi = d(\vartheta E + \log \Phi).$$

Thus we see that the quantity

$$\vartheta(dE - \overline{\delta A})$$

is the total differential of a certain thermodynamic function. The above result really contains the second law of thermodynamics. In fact, in the classical presentation of thermodynamics the quantity dE is the sum of the work $\overline{\delta A}$ done by the external forces, and the "amount of heat" δQ received by the system during the elementary transition. Since the quantity δQ is formally *defined* as the difference $dE - \overline{\delta A}$, it is clear that it need not necessarily be the total differential of some thermodynamic function. However, the second law of thermodynamics tells us that the quantity $\delta Q/T$, where T is the absolute temperature of the system, is always a total differential.

The most satisfactory statement of this law is as follows: there exists such a function ϑ of the temperature, and such a function W of the temperature and of the external parameters that for any elementary change of the thermodynamic state of a system we have

$$\vartheta \delta Q = dW.$$

In other words the function ϑ represents the integrating factor of the quantity δQ.

The existence of such an integrating factor, depending only on temperature, represents one of the formulations of the second law of the thermodynamics. This is, however, exactly what we have proved if one considers the parameter ϑ to be directly connected with the temperature of the system.

In classical thermodynamics the absolute temperature T is *defined* by the relation

$$\vartheta = \frac{1}{kT} \qquad (k = \text{Boltzmann's constant}) \tag{81}$$

in terms of this integrating factor ϑ, the existence of which is postulated in the second law. Introducing

$$kW = k[E\vartheta + \log \Phi(\vartheta)] = k\left(\log \Phi(\vartheta) - \vartheta \frac{\partial \log \Phi}{\partial \vartheta}\right) = S$$

we can write the second law of thermodynamics in the form

$$\frac{\delta Q}{T} = dS.$$

The thermodynamic function S is known as the entropy of the system. The above given argument is the complete foundation of the second law of thermodynamics in the frame of our theory, and indicates the reasonableness of the relation (81) as a universal postulate pertaining to any system of the assumed type.

Example: for an ideal monotomic gas we have, according to chapter V,

$$E = -\frac{\partial \log \Phi}{\partial \vartheta} = \frac{3n}{2\vartheta}, \qquad dE = -\frac{3n}{2\vartheta^2}\,d\vartheta,$$

$$-\overline{\delta A} = p\,dV = \frac{n}{\vartheta V}\,dV$$

so that

$$\delta Q = dE - \overline{\delta A} = -\frac{3n}{2\vartheta^2}\, d\vartheta + \frac{n}{\vartheta V}\, dV = \frac{3kn}{2}\, dT + \frac{knT}{V}\, dV,$$

$$\frac{\delta Q}{T} = \frac{3kn}{2}\frac{dT}{T} + kn\,\frac{dV}{V} = d\left(\frac{3}{2}\, kn \log T + kn \log V\right) = dS,$$

from which follows

$$S = \frac{3}{2}\, kn \log T + kn \log V + C, \tag{82}$$

where C is a constant.

As we have already mentioned, we will not be concerned in this book with the derivation of the entire system of thermodynamics; this can be done on the basis of its two fundamental laws without relation to the statistical point of view. Our task was only to show that these two fundamental laws represent the necessary consequences of our point of view.

However, in concluding this chapter, we must discuss in some detail a number of fundamental questions connected with the notion of entropy.

33. The properties of entropy. The notion of entropy is one of the most important physical notions from a theoretical as well as from a practical point of view. Very few other notions can compete with it in respect to the abundance of attempts to clarify its theoretical and philosophical meaning.

Many of these attempts are closely connected with the statistical interpretation of the phenomenon of heat, and are sometimes directly based on such an interpretation. Our problem is to see to what extent such probabilistic foundations of thermodynamics give a basis for certain far reaching statements concerning the nature of entropy.

In the above discussions we have always defined the quantity ϑ as a single root of the equation

$$\frac{d \log \Phi(\alpha)}{d\alpha} + E = 0.$$

Hence, we have seen in chapter IV, ϑ coincides with the value of the variable α for which the function

$$e^{E\alpha}\Phi(\alpha)$$

and, consequently, also its logarithm

$$E\alpha + \log \Phi(\alpha) \tag{83}$$

possess a single minimum. But, by the definition in the previous section the quantity

$$E\vartheta + \log \Phi(\vartheta)$$

is equal to the entropy of the system except for a constant factor. Thus we see that the entropy can be defined (up to a constant factor) as the minimum value of the function (83) with the argument ϑ. This permits us to establish one important property of the entropy.

Suppose we have two systems which together form a composite system. We will use the indices 1 and 2 for quantities pertaining to these systems, and no index for the composite system. Since $E = E_1 + E_2$ and

$$\Phi(\alpha) = \Phi_1(\alpha)\Phi_2(\alpha)$$

we have

$$E\alpha + \log \Phi(\alpha) = E_1\alpha + \log \Phi_1(\alpha) + E_2\alpha + \log \Phi_2(\alpha),$$

so that

$$\begin{aligned} S &= k[E\vartheta + \log \Phi(\vartheta)] \\ &= k[E_1\vartheta + \log \Phi_1(\vartheta)] + k[E_2\vartheta + \log \Phi_2(\vartheta)]. \end{aligned}$$

Since the functions $E_1\alpha + \log \Phi_1(\alpha)$ and $E_2\alpha + \log \Phi_2(\alpha)$ reach a minimum for $\alpha = \vartheta_1$ and $\alpha = \vartheta_2$ respectively, we obtain $S \geq k[E_1\vartheta_1 + \log \Phi_1(\vartheta_1)] + k[E_2\vartheta_2 + \log \Phi_2(\vartheta_2)] = S_1 + S_2$. This means that the entropy of the system obtained by bringing into thermal interaction two previously isolated systems is never smaller than the sum of the entropies of the two components; the two quantities become equal if the two components have originally the same temperature.

It must be noticed here that this theorem is often used, without sufficient foundation, in reaching rather broad conclusions, and the theorem itself is often expressed in rather indefinite and exaggerated terms. For instance, one states that because of thermal interaction of material bodies the entropy of the universe is constantly increasing. It is also stated that the entropy of a system "which is left to itself" must always increase; taking into account the probabilistic foundation of thermodynamics, one often ascribes to this statement a statistical rather than an absolute character. This formulation is wrong if only because the entropy of an isolated system is a thermodynamic function—not a phase-function—which means that it cannot be considered as a random quantity; if E and all λ_s remain constant the entropy cannot change its value whereas by changing these parameters in an appropriate way we can make the entropy increase or decrease at will. Some authors[1] try to generalize the notion of entropy by considering it as being a phase function which, depending on the phase, can assume different values for the same set of thermodynamical parameters, and try to prove that entropy so defined must increase, with overwhelming probability. However, such a proof has not yet been given, and it is not at all clear how such an artificial generalization of the notion of entropy could be useful to the science of thermodynamics.

We will arrive at a much more rational formulation of the problem if we will consider the given system as a part of another more extensive system. Let us assume that this more extensive system (which we will characterize by the asterisk) is in thermal equilibrium (compare chapter V, section 25). In other words, our system represents only an infinitesimally small part of this large system. In this case the energy E of the given system is no longer determined by the values of the parameters ϑ, $\lambda_s(\vartheta = \vartheta^*$ since, being in thermal equilibrium with the

[1]Comp. Borel, Mécanique statistique classique, Paris 1925.

larger system, our system has the same temperature) but is a random quantity whose distribution law is given approximately (compare chapter V, section 20) by the density:

$$\frac{\Omega(E)e^{-\vartheta E}}{\Phi(\vartheta)}.$$

It is clear that the relation $E = -(\partial \log \Phi)/\partial\vartheta$ cannot hold in this case since the left hand side of this expression is a random quantity, whereas the right hand side is determined exactly by the temperature ϑ^* of the thermostat. Thus the quantity E can no longer play the same role as before with respect to the second law of the thermodynamics.

However, the mean value of E, given by

$$\overline{E} = \int \frac{E\Omega(E)e^{-\vartheta E}}{\Phi(\vartheta)}\,dE = -\frac{\partial \log \Phi}{\partial \vartheta},$$

is related to ϑ in the same way as in the case of an isolated system. Thus, defining the entropy of the system by the expression

$$\overline{S} = k[\overline{E}\vartheta + \log \Phi(\vartheta)] \tag{84}$$

we deal with a thermodynamic function which is subject to all the arguments set forth in the previous section; in particular the proof of the second law of thermodynamics remains completely unchanged.

Using the formula (84) one can give a simple derivation of one of the most fundamental inequalities of thermodynamics. Let us assume that the system described by the numbers E_1, ϑ_1, S_1, interacts thermally with another system described by the numbers E_2, ϑ_2, S_2 (this second system is necessarily considered to be a thermostat). Let the numbers $E = E_1 + E_2$, ϑ, S characterize this composite system. Since the function $E_1\alpha + \log \Phi_1(\alpha)$ possesses a minimum for $\alpha = \vartheta_1$ we have

$$S_1 = k[E_1\vartheta_1 + \log \Phi_1(\vartheta_1)] \leq k[E_1\vartheta + \log \Phi_1(\vartheta)].$$

On the other hand formula (84) gives us

$$\overline{S}_1 = k[\overline{E}_1\vartheta + \log \Phi_1(\vartheta)],$$

where $\bar{E}_1$ is the mean energy and $\bar{S}_1$ the entropy of the given system considered as a part of the composite system. It follows:

$$\bar{S}_1 - S_1 \geq k\vartheta(\bar{E}_1 - E_1)$$

where, because of the approximate expression (51),

$$\bar{E}_1 \approx -\frac{d \log \Phi_1}{d\vartheta}, \qquad E_1 = -\frac{d \log \Phi_1}{d\vartheta_1}.$$

If $\vartheta_1 \leq \vartheta \leq \vartheta_2$, we have $E_1 \geq \bar{E}_1$ since, as we know, the function $(d \log \Phi)/d\alpha$ never decreases. Therefore:

$$\vartheta(\bar{E}_1 - E_1) \geq \vartheta_2(\bar{E}_1 - E_1).$$

The same relation takes place for $\vartheta_1 \geq \vartheta \geq \vartheta_2$ since in this case $E_1 \leq \bar{E}_1$.

Thus, in all cases we have

$$\bar{S}_1 - S_1 \geq k\vartheta_2(\bar{E}_1 - E_1) = \frac{\bar{E}_1 - E_1}{T_2}$$

which can be described by the following statement. *The entropy increase of a system resulting from its thermal interaction with any other system cannot be smaller than the energy increase of the first system divided by the absolute temperature of the second.*

One can, of course, generalize the notion of the entropy by writing for any state of the system

$$S = k[E\vartheta + \log \Phi(\vartheta)]$$

in which case the entropy itself becomes a random quantity.

For such a definition of entropy the second law of thermodynamics loses meaning, remaining applicable, however, to the mean value $\bar{S}$ which apparently is identical with the expression (84). In this case the distribution law of the given system in its phase space is given by the probability density

$$q = \frac{e^{-E\vartheta}}{\Phi(\vartheta)} = e^{-S/k}$$

from which follows

$$S = -k \log q.$$

This expression is often used to justify the statement that "the entropy of a system is proportional to the logarithm of the probability of the corresponding state" (Boltzmann's postulate). This statement, which is absolutely meaningless in the case of an isolated system, obtains, as we see, some meaning for a system in the larger system. This can be accomplished however, only by using the above described generalization of the notion of entropy which is introduced "ad hoc". In fact, one must not forget that this notion is used in connection with the second law of thermodynamics which loses meaning when the generalized definition of entropy is used. All existing attempts to give a general proof of this postulate must be considered as an aggregate of logical and mathematical errors superimposed on a general confusion in the definition of the basic quantities.[2] In the most serious treatises on that subject (for example: R. H. Fowler "Statistical Mechanics" Cambridge 1936) the authors refuse to accept this postulate, indicating that it cannot be proved, and cannot be given a sensible formulation even on the basis of the exact notions of thermodynamics.

However, proceeding in this direction we can obtain some reasonable and rather interesting results.

Consider two isolated systems, characterized by the indices 1 and 2, which form a composite system (characterized by no indices).

The total energy E of the composite system can be distributed in many different ways between the two thermally interacting components (the probabilities of various distributions have been considered in detail in section 22, chapter V). Let us write $p(E_1)\, dE_1$ for the probability that the energy of the first system lies in the interval $[E_1, E_1 + dE_1]$.

We know that the sum of the entropies $S_1 + S_2$ of the isolated systems can never exceed the entropy S of the composite system so that

---

[2]Comp. the "proof" in "Thermodynamik" by M. Planck, and the corresponding critique in "Statistical Mechanics" by R. H. Fowler.

$$S - (S_1 + S_2) \geq 0.$$

It can be shown that the excess is proportional, except for a constant additive term, to the logarithm of the quantity $p(E_1)$. This statement has the following meaning. Consider two systems which interact thermally forming a single system with the energy E and the entropy S. At some moment we isolate the two systems from each other thus obtaining two systems with energies E_1 and $E_2 (E_1 + E_2 = E)$, and the entropies S_1 and $S_2 (S_1 + S_2 \leq S)$. The quantities E_1 and E_2 must be considered as random quantities with the probability densities $p(E_1)$ and $p(E_2)$. Our statement tells us that the quantity log p(E_1) (as well as $\log p(E_2)$) is connected linearly with the quantity $S - (S_1 + S_2)$.

To prove this we must remember that, according to formula (27) chapter IV:

$$p(E_1) = \frac{\Omega_1(E_1)\,\Omega_2(E_2)}{\Omega(E)}.$$

On the other hand, because of (42) chapter V, we have, approximately,

$$\Omega(E) = \frac{e^{E\vartheta}\Phi(\vartheta)}{(2\pi B)^{1/2}} = \frac{e^{S/k}}{(2\pi B)^{1/2}}\,.$$

In similar fashion

$$\Omega_1(E_1) = \frac{e^{S_1/k}}{(2\pi B_1)^{1/2}}\,, \qquad \Omega_2(E_2) = \frac{e^{S_2/k}}{(2\pi B_2)^{1/2}}$$

where

$$B = \left(\frac{d^2 \log \Phi(\alpha)}{d\alpha^2}\right)_{\alpha=\vartheta}, \quad B_1 = \left(\frac{d^2 \log \Phi_1(\alpha)}{d\alpha^2}\right)_{\alpha=\vartheta_1},$$

$$B_2 = \left(\frac{d^2 \log \Phi_2(\alpha)}{d\alpha^2}\right)_{\alpha=\vartheta_2}.$$

This gives us the approximate relation:

(85) $$\log p(E_1) = -\frac{1}{k}[S - (S_1 + S_2)] + \frac{1}{2}\log\frac{2\pi B}{B_1 B_2}.$$

The second term on the right hand side cannot be considered as a constant (independent of chance) quantity since B_1 and B_2 depend on ϑ_1 and ϑ_2 which are in turn connected with the random quantity E_1 . It can be easily shown, however, that in the case when E_1 does not deviate appreciably from its mean value $\overline{E}_1$ (for which $\vartheta_1 = \vartheta$), the values of B_1 and B_2 are very close to

$$\left(\frac{d^2 \log \Phi_1(\alpha)}{d\alpha^2}\right)_{\alpha=\vartheta}, \qquad \left(\frac{d^2 \log \Phi_2(\alpha)}{d\alpha^2}\right)_{\alpha=\vartheta}.$$

These quantities, which do not contain any random element will be denoted by B_1' and B_2' . We will indicate here only the main steps of this proof leaving the more detailed calculations to the reader.

Because of the fundamental law of composition of the generating functions, the logarithms of these functions, as well as all derivatives of these logarithms, are infinitely large quantities of the order n where n is the number of molecules forming the system. In particular the quantities E_1 , E_2 , B_1 , B_2 , B_1' , B_2' are infinitely large quantities in the above described sense. For the same reason the quantities $B_1 - B_1'$ and $B_2 - B_2'$ have orders of magnitude $n(\vartheta_1 - \vartheta)$, and $n(\vartheta_2 - \vartheta)$ for small values of $(\vartheta_1 - \vartheta)$ and $(\vartheta_2 - \vartheta)$. Thus the ratios B_1'/B_1 and B_2'/B_2 differ from unity by $(\vartheta_1 - \vartheta)$ and $(\vartheta_2 - \vartheta)$. On the other hand:

$$E_1 = -\left(\frac{d \log \Phi_1(\alpha)}{d\alpha}\right)_{\alpha=\vartheta_1}$$

and, approximately, (according to (51) chapter V):

$$\overline{E}_1 = -\left(\frac{d \log \Phi_1(\alpha)}{d\alpha}\right)_{\alpha=\vartheta}$$

so that

$$E_1 - \overline{E}_1 = -\left(\frac{d^2 \log \Phi_1(\alpha)}{d\alpha^2}\right)_{\alpha=\vartheta}(\vartheta_1 - \vartheta) + 0[n(\vartheta_1 - \vartheta)^2]$$

$$= B_1'(\vartheta_1 - \vartheta) + 0[n(\vartheta_1 - \vartheta)^2].$$

Thus we conclude that for

$$E_1 - \bar{E}_1 = 0(n)$$

the difference $\vartheta_1 - \vartheta$ is an infinitesimally small quantity of the order:

$$\frac{E_1 - \bar{E}_1}{n}.$$

Consequently:

$$\frac{B_1'}{B_1} = 1 + 0\left(\frac{E_1 - \bar{E}_1}{n}\right), \qquad \frac{B_2'}{B_2} = 1 + 0\left(\frac{E_2 - \bar{E}_2}{n}\right),$$

and

$$\log B_1' - \log B_1 = 0\left(\frac{E_1 - \bar{E}_1}{n}\right),$$

$$\log B_2' - \log B_2 = 0\left(\frac{E_2 - \bar{E}_2}{n}\right).$$

If, as we have assumed, the deviation of E_1 from its mean value $\bar{E}_1$ is negligibly small in comparison with n (which is extremely probable because, as we shall see later, the mean square deviation of the quantity E, is of the order of magnitude of $n^{1/2}$), the estimates given above permit us to substitute the quantities B_1' and B_2' for the quantities B_1 and B_2 in the formula (85), and obtain the approximate relation

$$\log p(E_1) \approx \frac{1}{k}(S_1 + S_2 - S) + \frac{1}{2} \log \frac{2\pi B}{B_1' B_2'}$$

which proves our statement.

34. Other thermodynamical functions. Before finishing this chapter we want to consider the expression of some other important thermodynamic functions in terms of our general theory.

1. *The thermodynamical potential or characteristic function of Planck*

$$\Psi(\vartheta, \lambda_s) = \log \Phi(\vartheta)$$

is a convenient function because of the fact that almost all most important thermodynamical functions of a system can be expressed in terms of it. Thus,

$$E = -\frac{\partial \Psi}{\partial \vartheta}, \qquad S = k\left(\Psi - \vartheta \frac{\partial \Psi}{\partial \vartheta}\right),$$

$$\overline{X}_s = -\frac{1}{\vartheta}\frac{\partial \Psi}{\partial \lambda_s} \qquad (s = 1, 2, \cdots, r),$$

etc.

2. *Heat capacity.* Let us assume that the temperature and the external parameters of a given system are subject simultaneously to some infinitesimal change. The heat capacity of the system is defined as the quantity

$$C = \frac{\delta Q}{dT} = \frac{dE}{dT} - \frac{\overline{\delta A}}{dT}$$

where $\overline{\delta A}$ is the mean value of the element of work discussed in section 30. It is clear that this quantity assumes different values for different changes of the parameters λ_s.

In the particular case when all the parameters λ_s remain constant ($d\lambda_s = 0$) we have

$$C = \frac{\partial E}{\partial T} = \frac{(\partial E/\partial \vartheta)\, d\vartheta}{d(1/k\vartheta)} = -k\vartheta^2 \frac{\partial E}{\partial \vartheta} = k\vartheta^2 \frac{\partial^2 \log \Phi}{\partial \vartheta^2} = k\vartheta^2 B.$$

Let us consider the special case of an ideal monatomic gas enclosed in a vessel, subject to no external forces except the reactions of the vessel's walls. In this case the only external parameter will be the volume V of the gas. Since $B = (3n)/(2\vartheta^2)$, the heat capacity for constant volume is given by

$$C_V = \frac{3}{2} kn. \tag{86}$$

Thus it is independent not only on the volume and temperature but also on the nature of the gas.

The heat capacity of the gas calculated under the assumption of constant pressure also plays an important role in physics.

$$p = nkT/V = \text{const.}$$

i.e.

$$dT = \frac{p}{nk}\,dV. \tag{87}$$

In order to calculate this heat capacity, we notice that the entropy of the ideal gas is given according to (82) by the expression:

$$S = k[E\vartheta + \log \Phi(\vartheta)] = kn \log V + \frac{3}{2}\,kn \log T + \text{const}.$$

Since, according to the second law of thermodynamics,

$$\delta Q/T = dS,$$

we conclude that for any changes of V and T the heat capacity is given by:

$$C = T\,\frac{dS}{dT} = knT\,\frac{dV}{V\,dT} + \frac{3}{2}\,kn$$

(in the particular case $dV = 0$ this reduces to the expression (86)). In the case of constant pressure dV and dT are related by (87) giving us the result: $C_p = (5/2)kn$ (remembering that $nkT/V = p$).

Thus the quantity $C_p/C_V = 5/3$ represents a "universal constant" which remains the same for any amount of any monatomic gas under any physical conditions. This statement is generally in a good agreement with experiments, and the observed deviations can always be satisfactorily explained by the fact that the actual gases are never ideal.

CHAPTER VIII

DISPERSION AND THE DISTRIBUTIONS OF SUM FUNCTIONS

35. The intermolecular correlation. Let us consider an isolated system consisting of a large number n of molecules, and let us select any two of these molecules. Let $\varphi(P)$ be a phase function of our system depending only on the dynamic coordinates of the first molecule, and $\psi(P)$ a phase function depending only on the dynamic coordinates of the second molecule. The functions $\varphi(P)$ and $\psi(P)$, considered as random quantities, are not statistically independent of one another since their dynamic coordinates are related by the condition that the total energy E of the system remains constant. Because of the large number of individual molecules composing the system one should expect, of course, that the dependence between the quantities $\varphi(P)$ and $\psi(P)$ is very weak. In particular, one can expect that in any calculation the coefficient of correlation between these two quantities is infinitesimally small. As we shall soon see this is actually true. However, in many problems (in particular in the calculation of the dispersion of sum functions) one must calculate sums of a very large number of such correlation coefficients; these sums often may apparently diverge, and thus cannot be neglected.[1] For this reason it is necessary to have at least an approximate expression for the intermolecular correlation coefficient. This question will be discussed in the present section.

It is clear that in order to obtain the asymptotic formula we must impose certain limitations on the structure functions $\omega_1(x)$ and $\omega_2(x)$ of the two molecules selected as well as on the

[1]This is particularly so in the case when, not being satisfied by estimates of the order of magnitude of the increase of dispersion, we want to obtain its asymptotic expression. This occurs, for example, in the comparison between theoretical and experimental results concerning fluctuations.

functions $\varphi(P)$ and $\psi(P)$. The point is, that we are looking for asymptotic formulae under the assumption that the number of molecules n as well as the total energy E of the system becomes infinitely large; it is clear without any calculation that an unduly rapid increase of the four above mentioned functions would disrupt their general asymptotic relations.

For our purposes it is quite sufficient to assume that each of these four functions increases less rapidly than Cx^r, where x is the energy (of the first or second molecule) corresponding to a given argument, and C and r are positive constants. This condition is actually satisfied in practically all problems of statistical mechanics.

We will use the following notations:

$\Omega(x)$ = structure function of the system,
$\Phi(\alpha)$ = generating function of the system,
$U(x)$ = conjugate distribution of the system,

$$A(=E) = -\left(\frac{d \log \Phi(\alpha)}{d\alpha}\right)_{\alpha=\vartheta}, \qquad B = \left(\frac{d^2 \log \Phi(\alpha)}{d\alpha^2}\right)_{\alpha=\vartheta}.$$

We will use letters with the indices 1, 2, and 12 to denote quantities for the system without the first molecule, without the second molecule, and without both of them respectively. We will further denote by e_1, $\varphi_1(\alpha)$, $\omega_1(x)$, γ_1, dv_1 the energy, the generating function, the structure function, the phase space, and the volume element in it. The same letters with the index 2 will denote the same quantities for the second molecule, whereas the index 12 will refer to the combination of the two molecules. Obviously:

$$e_{12} = e_1 + e_2, \qquad \varphi_{12}(\alpha) = \varphi_1(\alpha)\varphi_2(\alpha), \qquad dv_{12} = dv_1\, dv_2$$

and the space γ_{12} is the direct product of γ_1 and γ_2.

The correlation coefficient connecting the functions φ and ψ is determined by the expression

$$R(\varphi, \psi) = \frac{\overline{(\varphi - \bar{\varphi})(\psi - \bar{\psi})}}{(D\varphi\, D\psi)^{1/2}},$$

where D is the symbol of dispersion. For the denominator of the above expression we have (according to (26) chapter IV) the following basic expression:

$$\overline{(\varphi - \bar{\varphi})(\psi - \bar{\psi})}$$

$$= \int_{\gamma_{12}} [\varphi(P) - \bar{\varphi}][\psi(P) - \bar{\psi}] \frac{\Omega^{(12)}(E - e_1 - e_2)}{\Omega(E)} dv_{12} .$$

Expressing the structure function in terms of the conjugate distribution functions (by means of (34) chapter IV), and putting $\alpha = \vartheta$ we have:

$$R(\varphi, \psi) = \frac{1}{(D\varphi\, D\psi)^{1/2}} \int_{\gamma_{12}} [\varphi(P) - \bar{\varphi}][\psi(P) - \bar{\psi}] \frac{e^{-\vartheta(e_1+e_2)}}{\varphi_1(\vartheta)\varphi_2(\vartheta)} \tag{88}$$

$$\cdot L(e_1 , e_2)\, dv_{12}$$

where, for brevity

$$L(e_1 , e_2) = \frac{U^{(12)}(E - e_1 - e_2)}{U(E)}.$$

As usual, our aim is to secure an asymptotic expression for $R(\varphi, \psi)$ under the assumption $n \to \infty$.

We begin with some general considerations. First of all we can see from the approximate formula (39) (chapter V) that the quantity $L(e_1 , e_2)$ is bounded uniformly for any values of e_1 and e_2 when $n \to \infty$; in fact, the quantity $U(E)$ is asymptotically equal to $(2\pi B)^{-1/2}$ (i.e. is an infinitely small quantity of the order $n^{-1/2}$), whereas, because of the same formula, the order of magnitude of $U^{(12)}(E - e_1 - e_2)$ is in any case not lower than $n^{-1/2}$.

We will also notice that for any arbitrary constants a, b, c, d, and for the constant values $k \geq 0$, $l \geq 0$, we have for $n \to \infty$:

$$I_n = \int_{e_1 > \log^2 n} [a\varphi(P) + b]^k [ce_1 + d]^l e^{-\vartheta e_1}\, dv_1 = 0(n^{-q}),$$

(89)

$$\int_{e_2 > \log^2 n} [a\psi(P) + b]^k [ce_2 + d]^l e^{-\vartheta e_2}\, dv_2 = 0(n^{-q}),$$

where g is any real number. Furthermore, because of the above assumptions concerning the properties of $\varphi(P)$ and $\omega_1(x)$, we have, for example, for the first integral:

$$| I_n | < C \int_{e_1 > \log^2 n} e_1^{kr+l} e^{-\vartheta e_1} \, dv_1 = C \int_{\log^2 n}^{\infty} x^{kr+l} e^{-\vartheta x} \omega_1(x) \, dx$$

$$< C_1 \int_{\log^2 n}^{\infty} x^{kr+l+r} e^{-\vartheta x} \, dx < C_1 \int_{\log^2 n}^{\infty} e^{-(\vartheta/2)x} \, dx = 0(n^{-g})$$

where C and C_1 are positive constants.

Turning now to the derivation of the asymptotic formula for the quantity $R(\varphi, \psi)$, we use the formula (99) (see appendix) for an estimate of the quantities $U^{(12)}(E - e_1 - e_2)$ and $U(E)$. Also, using the arguments given at the end of the appendix we find that for

$$| e_1 - \bar{e}_1 | < \log^2 n, \qquad \text{and} \qquad | e_2 - \bar{e}_2 | < \log^2 n:$$

$$U^{(12)}(E - e_1 - e_2) = \frac{1}{[2\pi B^{(12)}]^{1/2}} \exp\left[-\frac{w^2}{2B} \right] + \frac{S_n + T_n w}{B^{5/2}} + 0\left(\frac{1 + | w |^3}{n^2} \right),$$

where we put for brevity:

$$e_1 + e_2 - \bar{e}_1 - \bar{e}_2 = w.$$

On the other hand we have:

$$U(E) = \frac{1}{(2\pi B)^{1/2}} + S_n B^{-5/2} + 0(n^{-2}).$$

Thus, in the part γ'_{12} of the space γ_{12} which is characterized by the inequalities $e_1 < \log^2 n$ and $e_2 < \log^2 n$, we have

$$L(e_1\,, e_2)$$

$$= \left[\left(\frac{B}{B^{(12)}} \right)^{1/2} \exp\left[-\frac{w^2}{2B} \right] + (2\pi)^{1/2} \frac{S_n + T_n w}{B^2} + 0\left(\frac{1 + | w |^3}{n^{3/2}} \right) \right] \Big/ [1 + (2\pi)^{1/2} S_n B^{-2} + 0(n^{-3/2})]$$

$$= \left[\left(\frac{B}{B^{(12)}}\right)^{1/2}\left(1 - \frac{w^2}{2B}\right) + (2\pi)^{1/2}S_nB^{-2} + (2\pi)^{1/2}T_nB^{-2}w\right.$$

$$\left. + 0\left(\frac{1 + |w|^3}{n^{3/2}}\right)\right] \Big/ [1 + (2\pi)^{1/2}S_nB^{-2} + 0(n^{-3/2})].$$

Noticing that

$$\left(\frac{B}{B^{(12)}}\right)^{1/2} = 1 + \frac{b_1 + b_2}{2B} + 0(n^{-2}),$$

and denoting for brevity

$$1 + (2\pi)^{1/2}S_nB^{-2} = \Lambda,$$

we find

$L(e_1, e_2)$

$$= \frac{\Lambda + \dfrac{b_1 + b_2}{2B} + (2\pi)^{1/2}T_nB^{-2}w - \dfrac{w^2}{2B} + 0\left(\dfrac{1 + |w|^3}{n^{3/2}}\right)}{\Lambda + 0(n^{-3/2})}$$

(90)

$$= 1 + \frac{b_1 + b_2}{2B} + (2\pi)^{1/2}T_nB^{-2}w - \frac{w^2}{2B} + 0\left(\frac{1 + |w|^3}{n^{3/2}}\right).$$

The right hand side of this equality has the form:

$$-\frac{(e_1 - \bar{e}_1)(e_2 - \bar{e}_2)}{B} + R + 0\left(\frac{1 + |w|^3}{n^{3/2}}\right)$$

where R is formed by the terms of the type

$$K(e_i - \bar{e}_i)^s$$

(K is constant, $i = 1$ or 2, $s = 0, 1$ or 2). But:

$$\int_{\gamma_{12}} (\varphi - \bar{\varphi})(\psi - \bar{\psi})e^{-\vartheta(e_1+e_2)}K(e_1 - \bar{e}_1)^s\, dv_{12}$$

$$= K\int_{\gamma_1} (\varphi - \bar{\varphi})(e_1 - \bar{e}_1)^s e^{-\vartheta e_1}\, dv_1 \int_{\gamma_2} (\psi - \bar{\psi})e^{-\vartheta e_2}\, dv_2.$$

According to formula (48) chapter V, the second integral on the right hand side is an infinitesimally small quantity of order not lower than n^{-1}. The first integral is of the same order of magnitude for $s = 0$, and remains bounded from above for $s = 1$ or 2. Therefore the terms with $s = 0$ (i.e. $1 + (b_1 + b_2)/2B$) in the expression for R, when integrated over the entire space γ_{12}, give a quantity of the order of magnitude $0(n^{-2})$, in the estimate of the integral (88). All other terms (i.e. those with $s = 1$ and $s = 2$) contain a constant (with respect to the variables of integration) factor of the order n^{-1}, giving after integration quantities of the order $0(n^{-2})$. Thus:

$$\int_{\gamma_{12}} (\varphi - \bar{\varphi})(\psi - \bar{\psi})e^{-\vartheta(e_1+e_2)}R\, dv_{12} = 0(n^{-2}).$$

Since, on the other hand, because of (89):

$$\int_{\gamma''_{12}} (\varphi - \bar{\varphi})(\psi - \bar{\psi})e^{-\vartheta(e_1+e_2)}R\, dv_{12} = 0(n^{-2})$$

where $\gamma''_{12} = \gamma_{12} - \gamma'_{12}$, is the part of the space γ_{12} determined by the condition

$$\max(e_1, e_2) \geq \log^2 n$$

we conclude that:

$$\int_{\gamma'_{12}} (\varphi - \bar{\varphi})(\psi - \bar{\psi})e^{-\vartheta(e_1+e_2)}R\, dv_{12} = 0(n^{-2}).$$

Denoting by Q the remaining terms in the formula (90), and by C and C_1 we have:

$$\left| \int_{\gamma'_{12}} (\varphi - \bar{\varphi})(\psi - \bar{\psi})e^{-\vartheta(e_1+e_2)}Q\, dv_{12} \right|$$

$$< Cn^{-3/2} \int_{\gamma_{12}} |\varphi - \bar{\varphi}|\,|\psi - \bar{\psi}|\,(1 + |w|^3)e^{-\vartheta(e_1+e_2)}\, dv_{12}$$

$$= C_1 n^{-3/2}.$$

Comparing the estimates so obtained we have:

$$\int_{\gamma'_{12}} (\varphi - \bar{\varphi})(\psi - \bar{\psi}) e^{-\vartheta(e_1+e_2)} \frac{U^{(12)}(E - e_1 - e_2)}{U(E)} dv_{12}$$

$$= - \int_{\gamma'_{12}} (\varphi - \bar{\varphi})(\psi - \bar{\psi}) \frac{(e_1 - \bar{e}_1)(e_2 - \bar{e}_2)}{B} e^{-\vartheta(e_1+e_2)} dv_{12}$$

$$+ 0(n^{-3/2}).$$

Since, as we have seen above, the ratio

$$\frac{U^{(12)}(E - e_1 - e_2)}{U(E)}$$

remains uniformly bounded in the entire space γ_{12} for $n \to \infty$, the estimate given by (89) permits us to integrate both parts of the above relation over the entire space γ_{12}. Therefore, using (88), we obtain:

$$R(\varphi, \psi) = - \frac{1}{B} \int_{\gamma_1} \frac{(\varphi - \bar{\varphi})(e_1 - \bar{e}_1)}{(D\varphi)^{1/2}} \frac{e^{-\vartheta e_1}}{\varphi_1(\vartheta)} dv_1$$

$$\cdot \int_{\gamma_2} \frac{(\psi - \bar{\psi})(e_2 - \bar{e}_2)}{(D\psi)^{1/2}} \frac{e^{-\vartheta e_2}}{\varphi_2(\vartheta)} dv_2 + 0(n^{-3/2}).$$

Using the formula (48) chapter V

$$\int_{\gamma_1} \varphi e_1 \frac{e^{-\vartheta e_1}}{\varphi_1(\vartheta)} dv_1 = \overline{\varphi e_1} + 0(n^{-1})$$

we can give similar estimates for other integrals, in which φe_1 is replaced by φ, e_1, ψe_2, ψ, e_2. This gives

$$R(\varphi, \psi) = - \frac{1}{B} \frac{(\overline{\varphi e_1} - \bar{\varphi} \cdot \bar{e}_1)(\overline{\psi e_2} - \bar{\psi} \cdot \bar{e}_2)}{(D\varphi\, D\psi)^{1/2}} + 0(n^{-3/2})$$

(91)

$$= - \frac{(b_1 b_2)^{1/2}}{B} R(\varphi, e_1) R(\psi, e_2) + 0(n^{-3/2}).$$

Both correlation coefficients entering into this expression connect two phase functions of the same molecule, and can be easily calculated if these functions are known. The basic value of the asymptotic relation which is obtained in this way is the fact that the correlation coefficient pertaining to the phase functions of two different molecules is, asymptotically, inversely proportional to B (being an infinitesimally small quantity of the order n^{-1}) for ever increasing n.

Let us now consider the expression (91) under various possible assumptions.

1. In the case when the two selected molecules have the same structure, and the functions φ and ψ are the same, we obtain (putting $b_1 = b_2 = b$):

$$R(\varphi, \varphi) = -\frac{b}{B} R^2(\varphi, e) + 0(n^{-3/2}),$$

where e stands for the energy of one of the selected molecules.

2. If $\varphi = \psi$, and all the molecules have the same structure, we have $B = nb$ and:

$$R(\varphi, \varphi) = -\frac{1}{n} R^2(\varphi, e) + 0(n^{-3/2}).$$

3. If $\varphi = e_1$ and $\psi = e_2$, the formula (91) gives us:

$$R(e_1, e_2) = -\frac{(b_1 b_2)^{1/2}}{B} + 0(n^{-3/2}).$$

The negative sign of the coefficient R in this last case could be foreseen; in fact, since the stochastic relation between the energies of two molecules is determined entirely by the condition that the total energy of all molecules is a constant, the decrease of the energy of any one molecule favors stochastically the increase of the energy of any other molecule and vice versa.

4. Finally if all the molecules have the same structure and $\varphi = e_1$ and $\psi = e_2$, we have:

$$R(e_1, e_2) = -\frac{1}{n} + 0(n^{-3/2}).$$

This result is trivial since in this case the formula:

$$0 = DE = D \sum_{k=1}^{n} e_k = \sum_{k=1}^{n} De_k + \sum_{i \neq k} (De_i\ De_k)^{1/2} R(e_i\,, e_k)$$

$$= nb + n(n - 1)bR(e_i\,, e_k) \qquad (i \neq k)$$

leads to the *exact* relation:

$$R(e_i\,, e_k) = -\frac{1}{n-1} \qquad (i \neq k).$$

36. Dispersion and distribution laws of the sum functions. As we have seen in section 13, chapter III, the estimates of dispersion of sum functions play an important role in the foundation of our entire theory. It is, in fact, the smallness of the mean square deviations of these functions which permits us to state that they assume values which are very close to their mean values.

In the terminology of the theory of probabilities this is equivalent to the statement that, for an infinitely large number of molecules, the sum functions are subject to the law of the large numbers. In view of the discussions of the previous paragraph, obtaining estimates of dispersion of the sum function does not present any essential difficulties.

Suppose we have a sum phase function:

$$f(P) = \sum_{i=1}^{n} f_i(P)$$

where each term is a function of the dynamical coordinates of one molecule only. The mean value of the function f is:

$$\bar{f} = \sum_{i=1}^{n} \bar{f}_i\,.$$

In the general case, when for increasing n the quantities $\bar{f}_i$ remain between two limits of the same sign, the above given mean value is an infinitely large quantity of the order n.

In the most common case, when all the molecules as well

as all f_i-functions, are identical, the quantity $\bar{f}$ is directly proportional to n.

Let us now estimate the dispersion of the function f. We have

$$Df = \overline{\{(f - \bar{f})^2\}} = \overline{\left[\left\{ \sum_{i=1}^{n} (f_i - \bar{f}_i)\right\}^2\right]}$$

$$= \sum_{i=1}^{n} \overline{\{(f_i - \bar{f}_i)^2\}} + \sum_{i \neq k} \overline{\{(f_i - \bar{f}_i)(f_k - \bar{f}_k)\}}$$

$$= \sum_{i=1}^{n} Df_i + \sum_{i \neq k} (Df_i \, Df_k)^{1/2} R(f_i , f_k).$$

Assuming the functions f_i are subject to the limitations introduced in the previous section, we obtain from the formula (91)

$$Df = \sum_{i=1}^{n} Df_i - \frac{1}{B} \sum_{i \neq k} (b_i b_k \, Df_i \, Df_k)^{1/2} R(e_i , f_i) R(e_k , f_k)$$

(92)

$$+ 0(n^{1/2})$$

since the number of the terms in the second sum on the right hand side is of the order of n^2 (in the above expression we denoted by e_i the energy of that molecule related to the function f_i , and by b_i the dispersion of this energy). The above relation indicates that, under the assumed conditions:

$$Df = 0(n)$$

meaning that the mean square deviation of the function f is of order not greater than $n^{1/2}$ (i.e. considerably lower than the mean value of that function). This fact establishes the "representability" of the mean values of the sum functions, and permits us to identify them with the time averages which represent the direct results of any physical measurement.

Let us consider, as an example, the energy of a large component of some given system, and let us assume that this component contains the molecules with numbers from 1 to $n_1(<n)$. We have $f_i = e_i$ $(1 \leq i \leq n_1)$, $f_i = 0$ $(i > n_1)$,

$f = \sum_{i=1}^{n_1} e_i$, and consequently $Df_i = b_i$ $(i \leq n_1)$ $Df_i = 0$ $(i > n_1)$, $R(e_i, f_i) = 1$. Formula (92) now yields

$$Df = \sum_{i=1}^{n_1} b_i - \frac{1}{B} \sum_{\substack{i,k \leq n_1 \\ i \neq k}} b_i b_k + 0(n^{1/2}).$$

Putting $\sum_{i=1}^{n_1} b_i = B'$, we can rewrite the above in the form

$$Df = B' - \frac{1}{B}\left(B'^2 - \sum_{i=1}^{n_1} b_i^2\right) + 0(n^{1/2})$$

or, noticing that

$$\frac{1}{B} \sum_{i=1}^{n_1} b_i^2 = 0(1)$$

in the form

$$Df = \frac{B'(B - B')}{B} + 0(n^{1/2}).$$

The main term in this expression is just the dispersion for the Gaussian distribution, representing, approximately, the energy distribution of the large component (comp. Chapter V, section 22).

Let us also notice that in the most common case when all the molecules and all the functions f_i are identical ($f_i(P) = \psi(P)$, $1 \leq i \leq n$) the formula (92) gives us:

$$Df = n\, D\psi - \frac{1}{n} D\psi\, R^2(\psi, e)(n^2 - n) + 0(n^{1/2})$$

(93)

$$= n\, D\psi\, [1 - R^2(\psi, e)] + 0(n^{1/2}),$$

where e stands for the energy of a single molecule.

Let us turn now to a question concerning the distributions of sum functions. As usual, we will consider this as a limit problem, studying the form of the laws for $n \to \infty$.

We can expect without detailed calculation that we will encounter here the same Gaussian distribution as in the particular case of the energy distribution of a large component of a given system (comp. section 22, chapter V). In fact, since any sum function represents the sum of an infinitely large number of random quantities, the interrelation of these quantities is determined by the condition that the sum of the energies of individual molecules is equal to the total energy E of the system.

With an ever increasing number of molecules, the correlation between the dynamical coordinates of any two of them becomes very weak; we have seen, in fact, that the correlation coefficient of two molecules tends to zero when $n \to \infty$. Hence using a well known theorem of the theory of probability one can expect that the distribution of the sum functions for a large number of molecules will be, as a rule, similar to the Gaussian distribution.

Let us consider the sum function $F = \sum_{i=1}^{n} f_i(q_i)$ where q_i represents the set of the dynamic coordinates of that molecule associated with the function f_i. Let $v_i(x, y)$ be the volume of that part of the phase space for this molecule, where $e_i < x$, $f_i(q_i) < y$. Also put

$$\omega_i(x, y) = \frac{\partial^2 v_i(x, y)}{\partial x \, \partial y}.$$

We will denote by $V(x, y)$ and $\Omega(x, y)$ the functions which are determined in similar manner for the entire system (the functions f_i will, of course, be replaced by F). If we divide this system into two components (characterized by the indices 1 and 2) one can easily see that

$$V(a, b) = \int_{\substack{E<a \\ F<b}} dV = \int_{\Gamma_1} dV_1 \int_{\substack{E_2<a-E_1 \\ F_2<b-F_1}} dV_2$$

$$= \int_{\Gamma_1} V_2(a - E_1, b - F_1) \, dV_1$$

$$= \iint V_2(a - x, b - y) \Omega_1(x, y) \, dx \, dy$$

where the double integral is extended over the entire surface $(x\ y)$.

This gives us

$$(94) \qquad \Omega(a, b) = \frac{\partial^2 V}{\partial a\, \partial b} = \iint \Omega_1(x, y)\, \Omega_2(a - x, b - y)\, dx\, dy.$$

This formula, which is analogous to the law of composition for structure functions (20), can be easily extended to an arbitrary number of components.

Let us now try to express the distribution law of F in terms of the function $\Omega(x, y)$. It can be done most simply by referring to the original geometrical meaning of the probability. We have previously defined the probability of the relation

$$(95) \qquad b < F < b + \Delta b.$$

We are now interested in this as the area of that part of the energy surface where this relation is fulfilled approaches zero. For this purpose we have selected the particular surface metrics for which the measure of any region of the surface Σ_a is equal to the limit (for $\Delta a \to 0$) of a ratio in which the numerator is the volume measure of the layer $a < E < a + \Delta a$ located above the given surface, and denominator is equal to Δa. In particular, the measure $\Omega(a)$ of the entire energy surface is equal to $V'(a)$.

It follows that the probability of the relation (95) can be determined geometrically as the limit (for $\Delta a \to 0$) of the ratio of the volume of phase space for which $a < E < a + \Delta a$ and $b < F < b + \Delta b$, to the volume of the layer $a < E < a + \Delta a$.

But the first of these two volumes is given by

$$V(a + \Delta a, b + \Delta b) - V(a + \Delta a, b) - V(a, b + \Delta b) + V(a, b),$$

whereas the second is

$$V(a + \Delta a) - V(a).$$

Dividing the numerator and denominator of this ratio by Δa, we find that the probability in question is

$$\lim_{\Delta a \to 0} \{[V(a + \Delta a, b + \Delta b) - V(a + \Delta a, b) - V(a, b + \Delta b) + V(a, b)]/\Delta a\}/\{[V(a + \Delta a) - V(a)]/\Delta a\}.$$

The probability density $p(a, b)$ is now obtained by dividing the above expression by Δb and putting $\Delta b \to 0$. This yields

$$p(a, b) = \frac{\dfrac{\partial^2 V(a, b)}{\partial a\, \partial b}}{\dfrac{dV(a)}{da}} = \frac{\Omega(a, b)}{\Omega(a)}.$$

Starting from this basic formula, we will now obtain from it asymptotic formulae which are more convenient in further calculations. We will do this by using the composition law (94) for the function $\Omega(x, y)$ in the same manner as for the structure functions in chapter V.

Let us assume that our system consists of a large number n of molecules, and let us put, as usual

$$\varphi_i(\vartheta) = \int \omega_i(x) e^{-\vartheta x}\, dx \qquad (1 \leq i \leq n),$$

$$\Phi(\vartheta) = \prod_{i=1}^{n} \varphi_i(\vartheta) = \int \Omega(x) e^{-\vartheta x}\, dx$$

where ϑ is determined by the relation $(d \log \Phi)/d\vartheta + E = 0$.

Let us also denote

(96)
$$\frac{\omega_i(x, y) e^{-\vartheta x}}{\varphi_i(\vartheta)} = u_i(x, y) \qquad (1 \leq i \leq n),$$

$$\frac{\Omega(x, y) e^{-\vartheta x}}{\Phi(\vartheta)} = U(x, y).$$

Since

$$\int \omega_i(x, y)\, dy = \omega_i(x) \qquad (1 \le i \le n), \tag{97}$$

$$\int \Omega(x, y)\, dy = \Omega(x),^{2}$$

it follows that the functions $u_i(x, y)$ and $U(x, y)$ represent the probability densities of some two-dimensional distribution (the analogue of the conjugate distributions).

Generalizing the composition law (94) for n-components we obtain

$$\Omega(a, b) = \int \left\{ \prod_{i=1}^{n-1} \omega_i(x_i, y_i)\, dx_i\, dy_i \right\} \omega_n\left(a - \sum_{i=1}^{n-1} x_i, b - \sum_{i=1}^{n-1} y_i\right).$$

Expressing the functions Ω and ω_i through the functions U and u_i, we can easily find from the formulae (96) that:

$$U(a, b) = \int \left\{ \prod_{i=1}^{n-1} u_i(x_i, y_i)\, dx_i\, dy_i \right\} u_n\left(a - \sum_{i=1}^{n-1} x_i, b - \sum_{i=1}^{n-1} y_i\right)$$

This relation shows that $U(x, y)$ is the probability density of the two-dimensional distribution for the sum of n mutually independent terms distributed according with the densities $u_i(x, y)$ $(1 \le i \le n)$. Therefore, making certain assumptions concerning the limiting behavior of the functions ω_i and f_i, we can use the two-dimensional central limit theorem.[3]

[2]This last relation can be obtained, for example, by differentiating the relation

$$V(x) = V(x, +\infty) - V(x, -\infty) = \int_{-\infty}^{+\infty} \frac{\partial V(x, y)}{\partial y}\, dy.$$

[3]Although the proof of this theorem for this particular case has never been published, we do not think it expedient to burden the present exposition by its detailed presentation. Although this proof is very long, it does not present any fundamental difficulties, and the reader can develop it along the lines given in the appendix.

For large n,

$$U(a, b) \approx \frac{1}{2\pi\Delta} \exp\left\{-\frac{1}{2\Delta^2}[B_2(a - A_1)^2 + A_2(b - B_1)^2 - 2C_2(a - A_1)(b - B_1)]\right\}$$

where

$$A_1 = \sum_{i=1}^{n} a_i \, , \; a_i = \iint xu_i(x, y)\, dx\, dy \qquad (1 \leq i \leq n),$$

$$B_1 = \sum_{i=1}^{n} b_i \, , \; b_i = \iint yu_i(x, y)\, dx\, dy \qquad (1 \leq i \leq n),$$

$$A_2 = \sum_{i=1}^{n} \iint (x - a_i)^2 u_i(x, y)\, dx\, dy,$$

$$B_2 = \sum_{i=1}^{n} \iint (y - b_i)^2 u_i(x, y)\, dx\, dy,$$

$$C_2 = \sum_{i=1}^{n} \iint (x - a_i)(y - b_i) u_i(x, y)\, dx\, dy,$$

$$\Delta^2 = A_2 B_2 - C_2^2 \, .$$ [4]

Because of the relations (96), (97):

$$A_1 = \sum_{i=1}^{n} \frac{1}{\varphi_i(\vartheta)} \iint x\omega_i(x, y) e^{-\vartheta x}\, dx\, dy$$

$$= \sum_{i=1}^{n} \frac{1}{\varphi_i(\vartheta)} \int x\omega_i(x) e^{-\vartheta x}\, dx$$

$$= -\sum_{i=1}^{n} \frac{\varphi_i'(\vartheta)}{\varphi_i(\vartheta)} = -\frac{d \log \Phi(\vartheta)}{d\vartheta} = E$$

[4]To avoid any misunderstanding one should note that the notations used in this derivation do not correspond to those which we used in the one-dimensional case. Previously we used the letter A to symbolize moments of the first order and the letters B and C (with different indices) to symbolize the moments of the second order; now we use these indices to indicate the order of the moments.

therefore, for $a = E$,

$$U(E, b) \approx \frac{1}{2\pi\Delta} \exp\left[-\frac{A_2}{2\Delta^2}(b - B_1)^2\right]$$

and, because of (96),

$$\Omega(E, b) \approx \Phi(\vartheta)e^{E\vartheta} \cdot \frac{1}{2\pi\Delta} \exp\left[-\frac{A_2}{2\Delta^2}(b - B_1)^2\right].$$

Remembering that according to (42)

$$\Omega(E) \approx \frac{\Phi(\vartheta)e^{E\vartheta}}{(2\pi A_2)^{1/2}}$$

we finally find

$$p(E, b) = \frac{\Omega(E, b)}{\Omega(E)} \approx \frac{1}{[2\pi(\Delta^2/A_2)]^{1/2}} \exp\left[-\frac{A_2}{2\Delta^2}(b - B_1)^2\right].$$

In other words the limiting form is nothing but a Gaussian distribution with the center at B_1 and with the dispersion:

$$\frac{\Delta^2}{A_2} = B_2 - \frac{C_2^2}{A_2}.$$

It remains to clarify the meaning of these parameters, and to see whether their values coincide with those which we obtained before. For this purpose we will show, first of all, that $u_i(x, y)$ is the density of the two-dimensional probability distribution which governs (approximately) the pair of random quantities e_i , f_i . In fact, since the set of dynamic coordinates of the selected molecule is distributed in the phase space γ_i of this molecule according to the law characterized (approximately) by the density $(e^{-\vartheta e_i})/(\varphi_i(\vartheta))$, the probability that the inequalities $x < e_i < x + dx$ and $y < f_i < y + dy$ will be satisfied simultaneously is equal to the integral of the above expression, extended over that part of the phase space γ_i where the above inequalities are satisfied. Since, however, the integrated function can be considered as a constant (equal to $e^{-\vartheta x}/\varphi_i(\vartheta)$) in that part of the space, and since the volume of this region is given by $[\partial^2 V_i(x, y)]/[\partial x \, \partial y] \, dx \, dy = \omega_i(x, y) \, dx \, dy$, this probability becomes $(e^{-\vartheta x}/\varphi_i(\vartheta))\omega_i(x, y) dx \, dy = u_i(x, y) \, dx \, dy$. This proves our statement. From this it follows that

$$B_1 \approx \sum_{i=1}^{n} \bar{f}_i = \bar{F},$$

$$A_2 \approx \sum_{i=1}^{n} De_i \, , \, B_2 \approx \sum_{i=1}^{n} Df_i \, , \, C_2 \approx \sum_{i=1}^{n} (De_i \, Df_i)^{1/2} R(e_i \, , f_i).$$

Therefore, for the dispersion associated with the limiting form we get

$$(98) \quad B_2 - \frac{C_2^2}{A_2} \approx \sum_{i=1}^{n} Df_i - \frac{1}{A_2} \left\{ \sum_{i=1}^{n} (De_i \, Df_i)^{1/2} R(e_i \, , f_i) \right\}^2 .$$

This expression coincides with the formula (92), (within the limits specified in its derivation). The expression (98) also leads to the formula (93) if one assumes that all the functions f_i are identical, and all molecules have the same structure.

Let us also note that in the case when the functions f_i are not correlated with the energies e_i of the corresponding molecules (i.e. when all $R(e_i \, , f_i) = 0$) the expression (98) leads to the value $\sum_{i=1}^{n} Df_i$ for the dispersion of the function F. This fact could be foreseen before; in fact, we have already seen that the stochastic dependence between the dynamic coordinates of different molecules is due entirely to the condition $\sum_{i=1}^{n} e_i = E$. It is clear, therefore, that the functions of the dynamic coordinates of individual molecules not being correlated with the energies and therefore not being subject to the above conditions, must behave as non-correlated random quantities. Thus the dispersion of their sum must be equal to the sum of their dispersions.

A proof of the central limit theorem of the theory of probability. We find it necessary here to give a complete proof of the central limit theorem of the theory of probabilities, because that form of this proof which is most convenient for the purposes of statistical mechanics is somewhat different from the form usually encountered in mathematical texts. The point is, that in mathematics one naturally tends to formulate theorems in the most general way, sacrificing, thereby, considerations of the accuracy of the given estimates; in the case of the central limit theorem one tries to give a proof which would hold for the broadest possible class of initial distribution functions, without giving much attention to the smallness of the higher order terms. On the other hand in the case of the statistical mechanics we can limit ourselves to comparatively "smooth" distributions, paying more attention to a detailed estimate of the secondary terms.

This difference in the points of view results in a somewhat different treatment of the details of the proof, and prevents us from simply referring the reader to the standard mathematical texts. It may be noticed, however, that the general idea and the analytical method used in the proof remain essentially unchanged, so that the competent reader could actually do this himself.

The central limit theorem. Suppose we have a sequence of mutually independent random quantities which are governed by distribution functions with the probability densities $u_k(x)$ $(k = 1, 2, \cdots)$, *and let*

$$g_k(t) = \int e^{itx} u_k(x)\, dx \qquad (k = 1, 2, \cdots)$$

represent the characteristic functions corresponding to these distribution laws. Let us assume that:

1. *The functions* $u_k(x)$ *possess continuous derivatives, and there exists such a constant* A *that*

$$\int | u_k'(x) | \, dx < A \qquad (k = 1, 2, \cdots).$$

2. *The functions* $u_k(x)$ *possess finite moments of the first five orders which we will denote by* a_k, b_k, c_k, d_k, e_k. *Without restricting the generality of our proof we can put* $a_k = 0$ $(k = 1, 2, \cdots)$; *then there exist positive constants* α *and* β *such that:*

$$0 < \alpha < b_k < \beta, \bar{c}_k < \beta, d_k < \beta, \bar{e}_k < \beta \qquad (k = 1, 2, \cdots)$$

where $\bar{c}_k$ *and* $\bar{e}_k$ *represent the absolute moments of the third and fifth order of the functions* $u_k(x)$.

3. *There exist positive constants* a *and* b *such that for* $| t | < a$:

$$| g_k(t) | > b \qquad (k = 1, 2, \cdots).$$

4. *For each interval* (c_1, c_2) $(c_1 c_2 > 0)$ *there exists a number* $\rho(c_1, c_2) < 1$ *such that for any* t *within the interval* (c_1, c_2) *we have:*

$$| g_k(t) | < \rho \qquad (k = 1, 2, \cdots).$$

Let $U_n(x)$ *be the probability density of the sum of the first* n *terms in the given sequence of random quantities. Then, for* $n \to \infty$ *and* $| x | < 2 \log^2 n$, *we have*

$$U_n(x) = \frac{1}{(2\pi B_n)^{1/2}} \exp\left[-\frac{x^2}{2B_n}\right] + \frac{S_n + T_n x}{B_n^{5/2}} + 0\left(\frac{1 + | x |^3}{n^2}\right). \tag{99}$$

For any arbitrary x *we have:*

$$U_n(x) = \frac{1}{(2\pi B_n)^{1/2}} \exp\left[-\frac{x^2}{2B_n}\right] + 0\left(\frac{1}{n}\right)$$

where $B_n = \sum_{k=1}^{n} b_k$, and S_n *and* T_n *are quantities independent of* x, *increasing not faster than* n.

The proof. We start with the Maclaurin formula

$$g_k(t) = 1 - \frac{t^2}{2} b_k - \frac{it^3}{6} c_k + \frac{t^4}{24} d_k + \frac{t^5}{120} g_k^{(5)}(\theta_k t), \quad | \theta_k | < 1$$

the validity of which is due in this case to the fact that the existence of an absolute moment of a certain order implies the existence of the corresponding derivative of the characteristic function. Because of the postulate 2:

$$| g_k^{(5)}(\theta_k t) | \leq \int | x |^5 u_k(x)\, dx = \bar{e}_k < \beta \qquad (k = 1, 2, \cdots).$$

Denoting by $\gamma_k(t) = \log g_k(t)$ that branch of the logarithmic function which passes through zero for $t = 0$, we can easily prove that for $t \to 0$:

$$\text{(100)} \qquad \gamma_k(t) = -\frac{t^2}{2} b_k - \frac{it^3}{6} c_k + \frac{t^4}{8}\left(\frac{d_k}{3} - b_k^2\right) + 0(t^5)$$

which holds uniformly with respect to k. Let us put:

$$\sum_{k=1}^{n} b_k = B_n\,, \qquad \sum_{k=1}^{n} c_k = C_n\,, \qquad \sum_{k=1}^{n}\left(\frac{d_k}{3} - b_k^2\right) = D_n$$

and substitute in the previous formula $t = u/(B_n)^{1/2}$. We will always assume in future that $u = 0$ (log n). Taking the sum of (100) for k from 1 to n, we find:

$$\sum_{k=1}^{n} \gamma_k\left(\frac{u}{(B_n)^{1/2}}\right) = -\frac{u^2}{2} - \frac{iu^3}{6B_n^{3/2}} C_n + \frac{u^4}{8B_n^2} D_n + 0\left(\frac{u^5}{n^{3/2}}\right),$$

which can be reduced to

$$\prod_{k=1}^{n} g_k\left(\frac{u}{(B_n)^{1/2}}\right) = e^{-u^2/2}\left\{1 - \frac{iC_n}{6B_n^{3/2}} u^3 + \frac{D_n}{8B_n^2} u^4 - \frac{C_n^2}{72B_n^3} u^6 \right.$$

$$\left. + 0\left(\frac{| u^5 | + | u^9 |}{n^{3/2}}\right)\right\}.$$

We rewrite this relation as:

$$\prod_{k=1}^{n} g_k\left(\frac{u}{(B_n)^{1/2}}\right) = e^{-u^2/2}\left\{1 + iK_nB_n^{-3/2}u^3 + L_nB_n^{-2}u^4\right.$$

(101)

$$\left. - M_n^2B_n^{-3}u^6 + 0\left(\frac{|u^5| + |u^9|}{n^{3/2}}\right)\right\}$$

where (as always in the future) each capital letter with the index n denotes the sum of n real numbers, independent of u, which form bounded sets for $n \to \infty$.

As is well known from the general theory of characteristic functions

$$U_n(x) = \frac{1}{2\pi}\int e^{-itx}\left\{\prod_{k=1}^{n} g_k(t)\right\} dt.$$

We will divide the interval of integration into two parts, defined by $|t| < \log n/(B_n)^{1/2}$ and $|t| \geq \log n/(B_n)^{1/2}$, and will write I_1 and I_2 for the values of the corresponding partial integrals. In order to estimate the value of I_1 we will use a shorter form of the Maclaurin formula

$$g_k(t) = 1 - \frac{t^2}{2}b_k + g_k^{(3)}(\theta_k't)\frac{t^3}{6}, \qquad |\theta_k'| < 1.$$

Since

$$|g_k^{(3)}(\theta_k't)| \leq \int |x|^3u_k(x)\,dx = \bar{c}_k < \beta$$

we obtain

$$|g_k(t)| < \left|1 - \frac{t^2}{2}b_k\right| + \frac{\beta}{6}|t|^3.$$

Remembering that $1 - z \leq e^{-z}$ for any real z, and assuming $t^2 < 2/\beta < 2/b_k$ ($k = 1, 2, \cdots$), i.e. that $1 - t^2 b_k/2 > 0$ ($k = 1, 2, \cdots$) we have:

$$|g_k(t)| < \exp\left[-\frac{t^2}{2}b_k + \frac{\beta}{6}|t|^3\right] \qquad (k = 1, 2, \cdots).$$

Assuming further that $|t|$ is so small the second term in the exponent is smaller than half of the first term, we have:

$$|g_k(t)| < \exp\left[-\frac{t^2}{4}b_k\right].$$

Since the quantities b_k are bounded from below (postulate 2), we conclude that, for sufficiently small t, the above inequality will hold for any k; thus,

$$\left|\prod_{k=1}^{n} g_k(t)\right| < \exp(-t^2B_n/4).$$

Let this relation hold for $|t| < \delta$; then, for sufficiently large n,

$$\frac{\log n}{(B_n)^{1/2}} < \delta$$

and

$$\left|\int_{\log n/(B_n)^{1/2}}^{\delta} e^{-itx}\left\{\prod_{k=1}^{n} g_k(t)\right\} dt\right| < \int_{\log n/(B_n)^{1/2}}^{\infty} e^{-t^2B_n/4}\, dt$$

$$= \frac{1}{(B_n)^{1/2}}\int_{\log n}^{\infty} e^{-u^2/4}\, du = 0(\exp[-\tfrac{1}{4}\log^2 n]) = 0(n^{-2}).$$

For an estimate of other parts of the integral let us notice that because of the postulate 1

$$|g_k(t)| = \left|\int e^{itx}u_k(x)\, dx\right|$$

(103)

$$= \left|\left[\frac{e^{itx}}{it}u_k(x)\right]_{-\infty}^{+\infty} - \frac{1}{it}\int e^{itx}u_k'(x)\, dx\right| < \frac{A}{|t|},$$

since $u_k(+\infty) = u_k(-\infty) = 0$.[1] Because of the postulate 4

---

[1]The existence of the limits $u_k(+\infty)$ and $u_k(-\infty)$ follows from the integrability of the function $|u_k'(x)|$. The integrability of the function $u_k(x)$ itself leads us to the conclusion that in the limit the two above values must vanish.

there must exist a number ρ, $0 < \rho < 1$ such that for $\delta < t < 2A$

$$| g_k(t) | < \rho < 1 \qquad (k = 1, 2, \cdots)$$

so that

$$\left| \int_\delta^{2A} e^{-itx} \left\{ \prod_{k=1}^n g_k(t) \right\} dt \right| < \rho^n (2A - \delta) = 0(n^{-2})$$

whereas, because of (103) we have

$$\left| \int_{2A}^\infty e^{-itx} \left\{ \prod_{k=1}^n g_k(t) \right\} dt \right| < \int_{2A}^\infty \left(\frac{A}{t} \right)^n dt = 0(n^{-2}).$$

Since similar estimates can be obtained for corresponding parts of the region $t < 0$, the relation (102) in conjunction with the last two relations gives us

$$I_2 = 0(n^{-2})$$

which holds uniformly with respect to x.

Let us turn now to an asymptotic estimate of the integral

$$I_1 = \frac{1}{2\pi} \int_{|t| < (\log n)/(B_n)^{1/2}} e^{-itx} \left\{ \prod_{k=1}^n g_k(t) \right\} dt$$

$$= \frac{1}{2\pi (B_n)^{1/2}} \int_{-\log n}^{+\log n} \exp\left[- \frac{ixu}{(B_n)^{1/2}} \right] \left\{ \prod_{k=1}^n g_k\left(\frac{u}{(B_n)^{1/2}} \right) \right\} du.$$

Replacing the product under the integral by its asymptotic expression (101), we obtain

$$I_1 = \frac{1}{2\pi (B_n)^{1/2}} \int_{-\log n}^{\log n} \exp\left[- \frac{ixu}{(B_n)^{1/2}} - \frac{u^2}{2} \right] \left\{ 1 + iK_n B_n^{-3/2} u^3 \right.$$

$$\left. + L_n B_n^{-2} u^4 - M_n^2 B_n^{-3} u^6 + 0\left(\frac{| u^5 | + | u^9 |}{n^{3/2}} \right) \right\} du$$

$$= \frac{1}{2\pi (B_n)^{1/2}} \int_{-\log n}^{\log n} \exp\left[- \frac{ixu}{(B_n)^{1/2}} - \frac{u^2}{2} \right] du$$

$$+ \frac{iK_n}{2\pi B_n^2} \int_{-\log n}^{\log n} \exp\left[- \frac{ixu}{(B_n)^{1/2}} - \frac{u^2}{2} \right] u^3 \, du$$

$$+ \frac{L_n}{2\pi B_n^{5/2}} \int_{-\log n}^{\log n} \exp\left[- \frac{ixu}{(B_n)^{1/2}} - \frac{u^2}{2} \right] u^4 \, du$$

$$- \frac{M_n^2}{2\pi B_n^{7/2}} \int_{-\log n}^{\log n} \exp\left[- \frac{ixu}{(B_n)^{1/2}} - \frac{u^2}{2} \right] u^6 \, du + 0(n^{-2}).$$

We will denote the first four terms on the right hand side by A_1, A_2, A_3, A_4. Since $e^{-t^2/2}$ is known to be the characteristic function of the Gaussian distribution with center at zero and with dispersion 1, we have

$$\frac{1}{2\pi(B_n)^{1/2}} \int \exp\left[-iu \frac{x}{(B_n)^{1/2}} \right] \exp\left[- \frac{u^2}{2} \right] du$$

$$= \frac{1}{(2\pi B_n)^{1/2}} \exp\left[- \frac{x^2}{2B_n} \right]$$

and since

$$\left| \frac{1}{2\pi(B_n)^{1/2}} \int_{|u| > \log n} \exp\left[- \frac{iux}{(B_n)^{1/2}} - \frac{u^2}{2} \right] du \right|$$

$$\leq \frac{1}{\pi(B_n)^{1/2}} \int_{\log n}^{\infty} \exp\left[- \frac{u^2}{2} \right] du = 0(n^{-2})$$

we obtain finally

$$(104) \qquad A_1 = \frac{1}{(2\pi B_n)^{1/2}} \exp\left[- \frac{x^2}{2B_n} \right] + 0(n^{-2}).$$

For an estimate of the remaining three integrals we will introduce the notation

$$\int u^r e^{-u^2/2} \, du = m_r \qquad (r = 1, 2, \cdots).$$

Let us notice that, for sufficiently large u

$$u^r < e^{u^2/4}$$

so that for sufficiently large u

$$(105)\qquad \int_{|u|>\log n} u^r e^{-u^2/2}\,du < \int_{|u|>\log n} e^{-u^2/4}\,du$$

$$< \exp\left[-\tfrac{1}{4}\log^2 n\right] = 0(n^{-2}).$$

We will also assume that $|x| < 2\log^2 n$.

For an estimate of A_2 we notice that

$$\int_{-\log n}^{\log n} u^3 \exp\left[-\frac{ixu}{(B_n)^{1/2}} - \frac{u^2}{2}\right] du$$

$$= -i\int_{-\log n}^{\log n} u^3 \sin\frac{xu}{(B_n)^{1/2}} \exp\left[-\frac{u^2}{2}\right] du$$

$$= -\frac{ix}{(B_n)^{1/2}}\int_{-\log n}^{\log n} u^4 e^{-u^2/2}\,du + 0(x^3 n^{-3/2}).$$

Using the estimate (105), and substituting into A_2, we find

$$(106)\qquad A_2 = \frac{m_4 K_n x}{2\pi B_n^{5/2}} + 0\left(\frac{1 + |x|^3}{n^2}\right).$$

In similar manner the estimate of

$$\int_{-\log n}^{\log n} u^4 e^{-u^2/2} \cos\frac{xu}{(B_n)^{1/2}}\,du = m_4 + 0\left(\frac{1 + x^2}{n}\right)$$

gives us

$$(107)\qquad A_3 = \frac{m_4 L_n}{2\pi B_n^{5/2}} + 0\left(\frac{1 + x^2}{n^{5/2}}\right).$$

We also get

$$(108)\qquad A_4 = \frac{m_6 M_n^2}{2\pi B_n^{7/2}} + 0\left(\frac{1 + x^2}{n^{5/2}}\right).$$

Collecting the estimates (104), (106), (107), (108), and remembering that $I_2 = 0(n^{-2})$, we find

$$U_n(x) = \frac{1}{(2\pi B_n)^{1/2}} \exp\left[-\frac{x^2}{2B_n}\right]$$

(109)

$$+ \frac{m_4K_nx + m_4L_n + m_6\dfrac{M_n^2}{B_n}}{2\pi B_n^{5/2}} + 0\left(\frac{1 + |x|^3}{n^2}\right),$$

which proves the first part of our theorem. To prove the second part of the theorem it is sufficient to notice that the integrals in A_2 , A_3 and A_4 remain bounded uniformly with respect to $x(-\infty < x < +\infty)$ when $n \to \infty$, and that the estimates of I_1 and A_1 are also uniform with respect to x.

Remark. For many applications of the central limit theorem in the problems of statistical mechanics one often must use, together with the formula (109), a similar formula for $U_{n'}(x)$ where n' is so close to n that the difference $n' - n$ remains bounded for $n \to \infty$ (we often have simply $n' = n - 1$ or $n' = n - 2$). In these cases it is useful to remember that in writing the expression for $U_{n'}(x)$ it is not necessary to substitute n for n' in *all* capital letters on the right hand side of (109). In particular, it suffices to substitute $B_{n'}$ for B_n in the radical of the first term, leaving all other indices unchanged. Thus, for $|x| < 2\log^2 n$ we can write

$$U_{n'}(x) = \frac{\exp\left[-\dfrac{x^2}{2B_n}\right]}{(2\pi B_{n'})^{1/2}} + \frac{m_4K_nx + m_4L_n + m_6\dfrac{M_n}{B_n}}{2\pi B_n^{5/2}}$$

$$+ 0\left(\frac{1 + |x|^3}{n^2}\right).$$

In fact, a simple calculation, which we leave to the reader, shows that substituting n' for *all* indices n, we change our expression (because the limitation of $B_n - B_{n'}$, $K_n - K_{n'}$, $L_n - L_{n'}$, $M_n - M_{n'}$) only by an infinitesimally small quantity of an order of magnitude lower than that of the remaining term. Thus, we can use or omit such a substitution at our convenience.

NOTATIONS

NOTATIONS	Basic system G	Large component G_i	Small component g	Molecule g_i	Pair of molecules $g_i + g_k$	Complementary system $G - g_i$	Complementary system $G - g_i - g_k$	Large system
No. of molecules	n	n_i	—	1	2	$n - 1$	$n - 2$	—
Phase space	Γ	Γ_i	γ	γ_i	γ_{ik}	$\Gamma^{(i)}$	$\Gamma^{(ik)}$	—
The element of the phase space	dV	dV_i	dv	dv_i	dv_{ik}	$dV^{(i)}$	$dV^{(ik)}$	—
Energy	E	E_i	e	e_i	e_{ik}	$E^{(i)}$	$E^{(ik)}$	—
Region $E < x$	V_x	$(V_x)_i$	—	—	—	—	—	—
The volume $E < x$	$V(x)$	$V_i(x)$	$v(x)$	$v_i(x)$	$v_{ik}(x)$	$V^{(i)}(x)$	$V^{(ik)}(x)$	—

The surface $E = x$	Σ_x	—	—	—	—	—	—	—
Structure function	$\Omega(x)$	$\Omega_i(x)$	$\omega(x)$	$\omega_i(x)$	$\omega_{ik}(x)$	$\Omega^{(i)}(x)$	$\Omega^{(ik)}(x)$	—
Generating function	$\Phi(\alpha)$	$\Phi_i(\alpha)$	$\varphi(\alpha)$	$\varphi_i(\alpha)$	—	$\Phi^{(i)}(\alpha)$	—	$\Phi^*(\alpha)$
Conjugate distribution	$U^{(\alpha)}(x)$	$U_i^{(\alpha)}(x)$	$u^{(\alpha)}(x)$	$u_i^{(\alpha)}(x)$	—	—	—	—
Empirical Temperature	ϑ	ϑ_i	—	—	—	$\vartheta^{(i)}$	$\vartheta^{(ik)}$	ϑ^*
$-\left(\frac{d \log \Phi}{d\alpha}\right)_{\alpha=\vartheta}$	A, a	A_i	—	a_i	a_{ik}	$A^{(i)}$	$A^{(ik)}$	A^*
$\left(\frac{d^2 \log \Phi}{d\alpha^2}\right)_{\alpha=\vartheta}$	B	B_i	b	b_i	b_{ik}	$B^{(i)}$	$B^{(ik)}$	B^*
Entropy	S	S_i	—	—	—	—	—	—

INDEX